T4-ALD-119

PLATONISM AND CARTESIANISM IN THE PHILOSOPHY OF RALPH CUDWORTH

By

LYDIA GYSI

St. Mary's Abbey, West Malling, Kent/England

2

HERBERT LANG BERN

Printed in Switzerland by Stämpfli & Cie, Bern

ἂν οὖν τις ἔχῃ κάλλιον εἰπεῖν,
ἐκεῖνος οὐκ ἐχθρὸς ὢν
ἀλλὰ φίλος κρατεῖ.

Plato. Tim. 54a.

To

The Benedictine Community

at

St. Mary's Abbey, West Malling

FOREWORD

In recent years, when the arena of philosophical theology has been almost entirely occupied by the contending forces of existentialism and neoscholasticism, with logical empiricism as a somewhat cynical spectator, the Platonic element in Christian thought has tended either to be neglected as irrelevant or condemned as Christian only in name. In particular, that extremely interesting group of seventeenth-century thinkers who are known as the Cambridge Platonists, and of whom the subject of this book was perhaps the most distinguished member, have received little attention, though J. A. Passmore's study 'Ralph Cudworth, An Interpretation' must be mentioned as an honourable exception. Dr. Gysi's work is thus especially welcome and we can only regret that something like twelve years has had to elapse between its completion and its publication. It is an unashamedly professional study, though I do not think that any well educated Christian will find its perusal either difficult or unprofitable; its academic quality will be evident from its acceptance as a doctoral thesis in the Faculty of Philosophy of the University of Basle. It is however of much more than academic interest. That a work on a rather untypical Anglican philosopher and theologian should be produced by a Swiss Orthodox of protestant descent who, even while writing on Cudworth found her vocation to the religious life, and is now living in an Anglican Benedictine community as an Eastern Orthodox nun, provides a striking example of the way in which the thought and spirituality of the various Christian communions have intermingled to their mutual profit in the post-war decades. Nothing of the personal history of Dr. Lydia Gysi (in religion Mother Maria) is directly reflected in her study of Cudworth, but it will give it an added interest for all those who, like Cudworth himself, refuse to set up an impermeable barrier between the realm of thought and the things of the spirit. I read her manuscript with great profit shortly after it had been submitted for her doctorate and I am now very glad to have the opportunity of commending it to a wider public.

E. L. Mascall,

Professor of Historical Theology
in the University of London.

King's College, London.
October 1962.

LITERATURE

Works by Cudworth

The True Intellectual System of the Universe.

Original text, 1678. (Quoted as T. I. S.)

1743 Edited in two volumes with biographical notes by T. Birch.
1820 Newly edited in four volumes.
1845 Newly edited with an English translation of the Commentary of Professor Mosheim by J. Harrison.
1733 Edited in Jena in a Latin translation with Latin Commentary by Professor Mosheim.
1773 The same, newly edited in Leyden.
1706 Shortened English edition of the T. I. S. with alterations by Thomas Wise.

A Treatise concerning Eternal and Immutable Morality, edited posthumously in 1731 by E. Chandler, Bishop of Durham. *Quoted as Mor.*

A Treatise on Freewill (extract from the manuscripts of Cudworth). First edited with notes by John Allen. London 1838. *Quoted as Fr. W.*

Sermon before the House of Commons, with preface, on I John, 3, 4, 1647. *Quoted as 1st Sermon.*

The Victory of Christ, a sermon. I Cor. 15, 57. 2nd edition, 1670. *Quoted as 2nd Sermon.*

The Union of Christ and the Church, a Shadow. 1642.

A Discourse concerning the True Notion of the Lord's Supper. 1642.

Dantur Rationes Boni et Mali Aeternae et Indispensabiles.

Dantur Substantiae Incorporeae, Natura sua Immortales, 1651. (Edited as an appendix to A Treatise on Freewill.)

Extracts from the manuscripts of Cudworth by J. H. Muirhead in the "The Platonic Tradition in the Anglo-Saxon Philosophy", pp. 63 ff. *Quoted as Muirhead Manuscr.*

Works on Cudworth

F. J. Powicke	The Cambridge Platonists, a Study, 1926.
J. H. Muirhead	The Platonic Tradition in Anglo-Saxon Philosophy, 1931.
J. Martineau	Types of Ethical Theory. Vol. II, 1885.
C. Lowrey	The Philosophy of Ralph Cudworth, 1884.

W.R. Scott	An Introduction to Cudworth's "Treatise concerning Eternal and Immutable Morality" with Life and a Few Critical Notes, 1891.
E. Cassirer	Die Platonische Renaissance in England und die Schule von Cambridge, 1932.
G. v. Hertling	J. Locke und die Schule von Cambridge.
J. Tulloch	Rational Theology and Christian Philosophy in England in the Seventeenth Century, 1874.
Beyer	Ralph Cudworth als Ethiker, Staatsphilosoph und Ästhetiker (Diss.), 1935.
K. Schmitz	Cudworth und der Platonismus. Giessen. (Diss.), 1919.
G. Aspelin	Cudworth's Interpretation of Greek Philosophy. Göteborg, 1943.
R. Descartes	Meditations métaphysiques (avec Objections et Réponses): Discours de la Méthode. Regulae ad directionem ingenii. Principia Philosophiae. Lettres.

INDEX OF NAMES

Chapter I

THE THEORY OF SUBSTANCES

Cudworth originally planned a full account of the "true intellectual system of the universe" in three parts. It was to proceed by three big stages through the examination and refutation of three forms of determinism, the "mechanical atomical fate", the "pantheistic stoical fate" and the "theological Calvinist fate". He achieved only the first volume, and this contains a careful exposition of two different forms of atomism, both of which had already been developed in Greek philosophy and revived in his own time. The original and genuine form of atomism was part of the theistic philosophy of the Pythagoreans, and was renewed in a magnificent way by Descartes. This ancient atomism underwent a falsification in the philosophy of Democritus and Leukippus, who detached it from its theistic metaphysical foundation and turned it into a materialistic atheistic doctrine. In this debased form it was taken up and developed by Thomas Hobbes.

Cudworth himself adopted the Pythagorean atomism and praised the Cartesian theory of substances as its most conspicuous renewal[1], since it offers not only a strong defence against materialism, but also a solid foundation for a theistic philosophy[2]. In the field of Ontology, therefore, Cudworth regarded Descartes as a prominent ally, whose merits he highly esteemed.

[1] Mor. 301 ff. And here we can never sufficiently applaud that ancient atomic philosophy, so successfully revived of late by Cartesius, in that it shews distinctly what matter is, and what it can amount unto, namely nothing else but what may be produced from meer magnitude, figure, site, local motion, and rest.

[2] Ibid. 74. Whereas, if rightly understood it is the most impregnable bulwark against both (atheism and immorality); for this philosophy discovering that the ideas of sense are fantastical, must needs suppose another principle in us superior to sense, which judges what is absolutely and not fantastically or relatively only true and false.

It is fascinating to follow Cudworth's exposition of the theory of Substances. Almost imperceptibly the Cartesian scheme turns into something quite different, and in the end the dualism of substances is seen as a frail insignificant framework, which again and again is, as it were, burst from inside by the force of his thinking.

On the background of the enthusiastic approval Cudworth gave to Descartes' theory of substances, the refutation of the Cartesian principles of knowledge will appear all the more forcible. We therefore begin our research at the outskirts of Cudworth's thought, whence we shall slowly advance to its centre in order to find out his own lasting contribution to philosophy.

Cudworth distinguishes two principles of Being[1]: a corporeal and an incorporeal substance. He does not, however conceive of these as two inadequate conceptions of one infinite substance[2], but rather as intrinsically different and incommensurable. There is no immediate connection between them, nor any gradual transition from one into the other[3].

It is impossible to consider the two substances as infinite. Two infinities necessarily become finite by the fact of their being relative to each other.

The two substances have no necessary existence, nor do they subsist of themselves. In Cudworth's words they are not 'self-existent', but they have an ἀρχή, on which each depends in its own way. Cudworth does not discuss the question whether this ἀρχή is their first cause only, or both first cause and beginning with time[4]. The substances are created and finite[5]. Again, each in its own way exists and

[1] T. I. S. 159. The first general heads of all entity are:
a) resisting and antitypous extension
b) life: internal energy and self-activity
acting with express consciousness
or without express consciousness.

[2] Ibid. 830. Again that extension and life or cogitation are not two inadequate conceptions neither of one and the self same substance considered brokenly and by piecemeal.

[3] Ibid. 862. Neither can matter... efficiently produce Soul, any more than Soul matter.

[4] Ibid. 752.

[5] Ibid. 858. Though both matter and all imperfect souls and minds, were at first created by one perfect omnipotent understanding Being.

develops according to the principles of finite being, i.e. in space and time[1]. The corporeal substance has a direct relationship to space, it itself is space. Not so the incorporeal; it exists only indirectly in space through its vital union with the corporeal substance.

In the relationship to time we see a reverse order. The incorporeal substance can only actualise itself in time by progression both in Being and thought. The corporeal substance, on the other hand, exists in time only in so far as it is related to the incorporeal.

Both substances have in themselves virtual infinity[2], which in the corporeal substance is manifest in the capacity of indefinite increase, for we cannot imagine any space to which more space could not be added[3].

This "imitating infinity", as Cudworth calls it, shows itself in two ways in the incorporeal substance. When the incorporeal substance is activated as thought, the infinity is found in its "virtual omniformity" to all possible Being[4]. This is manifest in the fact that the souls can, in a discursive manner awaken in themselves all ideas. When the incorporeal substance acts as will, its virtual infinity is seen in the

[1] Ibid. 887. ... nothing which was properly made or created by God, and nothing which was not it self God, could be from eternity, or without beginning.

... Time was made together with the world, and there was no sooner or later, before time.

[2] Ibid. 647. And because infinity is perfection, therefore can nothing which includeth any thing of imperfection, in the very idea and essence of it, be ever truly and properly infinite; as numbers, corporeal magnitude, and successive duration. All which can only MENTIRI INFINITATEM counterfeit and imitate infinity, in their having more and more added to them infinitely, whereby notwithstanding they never reach it or overtake it.

[3] T. I. S. 644. How vast soever the finite world should be, yet is there a possibility of more and more magnitude and body, still to be added to it, further and further, by divine power, infinitely; or the world could never be made so great, no not by God himself, as that his own omnipotence could not make it yet greater.

Cf. ibid. 766 ... by space without the finite world, is to be understood, nothing but the possibility of body, further and further without end, yet so as never to reach to infinity: And such a space as this was there also, before this world was created, a possibility of so much body to be produced.

[4] Mor. 134. For the soul having an innate cognoscitive power universally (which is nothing else but a power of raising objective ideas within it self and intelligible reasons of any thing) it must needs be granted that it hath a potential omniformity in it.

capacity of transcending even a clear and distinct judgment of reason[1].

As substances, they cannot of themselves cease to be; they remain constant in the universe, the corporeal substance in quantity, the incorporeal in energy[2].

In the flux of Becoming it is only the accidents which change. In the corporeal substance the change consists in new relations between the atoms. In the incorporeal substance, when acting as thought, the change is found in new relations between the ideas; when acting as will, it is in the distribution of energy in its action "ad extra", as well as upon itself.

No new entity is created by either of these accidental changes. A fresh creative act of God would be needed for this[3].

If the substances are said to be to a certain degree indestructible, this does not imply that they have any guarantee of existence within themselves. As they are created by the will of God, so can they be at any time annihilated by this same will[4].

The two substances are in direct dependence on God who created them, but in no way co-ordinate. The corporeal substance is in a double dependence, firstly upon God, and secondly upon the incorporeal substance. The incorporeal substance has a priority and a "natural imperium" over the corporeal substance. This priority it undoubtedly has by nature, if not also in time in the sense that it

[1] Fr. W. 39. ...The HEGEMONIKON of the soul does sometimes extend itself further in way of assent than the necessary understanding goes, or beyond clear and distinct perceptions.

[2] Ibid. 30. As Cartesius supposes the same quantity of motion to be perpetually conserved in the universe... so more or less here and there, is the same stock of love and desire of good always alive, working in the soul...

[3] T. I. S. 739. Now no imperfect Being whatsoever, hath a sufficient emanative power to create any other substance or produce it out of nothing; the utmost that can be done by imperfect beings, is only to produce new accidents and modifications; as humane souls can produce new cogitations in themselves, and new local motion in bodies.

[4] Ibid. 868. Wherefore when it is said, that the immortality of the humane soul is demonstrable by natural reason, the meaning thereof is no more than this, that its substantiality is so demonstrable; from whence it follows that it will naturally no more perish or vanish into nothing, than the substance of matter it self: and not that it is impossible, either for it, or matter, by divine power to be annihilated.

was created first[1]. Since the substances are alien to each other, the "imperium" cannot be exerted directly, but requires mediating links. Cudworth's solution to the problem of this interaction between the two substances will show whether, and in what way, he has overcome the Cartesian dualism.

The Corporeal Substance

The corporeal substance manifests itself in particular bodies. It is divisible. Each body is therefore itself a manifestation of the corporeal substance; it is as many substances as it has parts[2].

Bodies are intelligible only according to their attributes of space. Cudworth thinks that we are justified in inferring from their intelligibility the objective reality of the bodies "without" the mind as Extension and its modifications. He regards this inference as an inference according to reason[3].

The attributes of Extension are magnitude, shape, place, motion and solidity of the particular part. These attributes are essential to the bodies, and inseparable from them. Changes of attributes do not affect the substance as such, they are natural to it. Body, as extension, is essentially "aliud extra aliud", alterity and exteriority[4].

[1] T. I. S. 844. So that cogitation is in order of nature before local motion and incorporeal before corporeal substance, the former having a natural imperium over the latter.

[2] Ibid. 829. ...no EXTENSUM whatsoever, of any sensible bigness, is truly and really, one substance, but a multitude or heap of substances, as many as there are parts into which it is divisible.

[3] Ibid. 28. ...when they further considered... the natural or corporeal principle, they being not able clearly to conceive any thing else in it, besides magnitude, figure, site and motion or rest, which are all several modes of extended bulk, concluded therefore according to reason, that there was really nothing else in bodies without, besides the various complexions and conjugations of those simple elements, that is, nothing but mechanism.

[4] Ibid. 829. RES EXTENSA... is extension, or distance, really existing without the mind... Now this in the nature of it, is nothing but ALIUD EXTRA ALIUD, ... and therefore perfect alterity, disunity and divisibility... Moreover one part of this magnitude, always standing without another, it is an essential property thereof to be antitypous or impenetrable.

We notice that Cudworth considers motion as a clearly and distinctly intelligible attribute. But we ask what reason has he for so doing? It is true that we can work out a motion from point to point, we can also conceive its whole tract. But motion is a continuum and as such cannot be comprehended by thought.

All other sensible qualities are intelligible only in so far as we can succeed in reducing them to atoms and their quantitative distribution[1]. As mere sensible qualities they are relative to the percipient, they are in fact no more than impressions on the soul without reality in the object "without". Their only objective reality lies in their being caused by the disposition of atoms[2]. Sensations, however, are not given to us in order to deceive us, but to give us joy, and to stimulate our mind to thought. A universe experienced only as Extension and its modifications would be a drab and dreary world, powerless to awaken and inspire our minds.

The corporeal substance has no power of action at all. It can act neither upon itself nor upon things without.

Because of its essential divisibility it continually disintegrates into indefinite multiplicity[3], towards both a maximum and a minimum; towards a maximum in its capacity of indefinite increase, towards a minimum in its divisibility ad indefinitum. This last, however, at the same time represents both minimum and maximum, minimum of extent and maximum of parts. In this fatal linear disintegration into the indefinite (*ἄπειρον*), the corporeal substance is always divorced from itself, unable to gather itself up into unity; it can, therefore, neither act upon itself, nor give itself a definite form. The only thing it can do, is passively to receive local motion from outside, and again

[1] Ibid. 29.

[2] Ibid. 33 ff. ...all phaenomena of inanimate bodies, and their various transformations, might be clearly resolved into these two things, partly something that is real and absolute in bodies themselves; and partly something that is phantastical in the sentient.

Ibid. 33. ...Yet they (fancies and passions) are wisely contriv'd by the author of nature, for the adorning and embellishing of the corporeal world to us.

[3] Ibid. 830. ...That which is wholly scattered out from itself into distance, and dispersed into infinite multiplicity.

passively to confer it to things outside[1]. Heterokinesis is pure passivity, which perforce leads to an indefinite regress. Yet by this its own impotence it points to a cause of a higher order which has in itself the power of autokinesis. Thus it indirectly testifies that local motion has its origin in an autokinetic, i.e. a cogitative, cause[2].

Motion is as equally natural and effortless to matter as rest. Deliverance from motion would bring no benefit but fruitless stagnation.

If on the other hand motion were not conferred from outside, but essential to matter itself, there could be no rest at all, and without rest no consistent form[3]. Consistency is the presupposition of all intelligibility. Matter as essentially motion could therefore not be conceived by thought at all.

We find here two different views side by side in Cudworth's thought. On the one hand he considers atoms as the smallest physical elements, thus denying their infinite divisibility. This assumption leads however to a difficulty. A smallest physical element would necessarily be simple; yet we have seen that even the smallest element is constituted by modifications of atoms. On the other hand Cudworth asserts the infinite divisibility of matter, and thereby denies that atoms are physical minima. Here he regards the theory of atomism as a mere explanatory hypothesis of the sensible world. He never tried to reconcile these two views[4].

[1] Ibid. 829. ...local motion to which it is also but passive, no body or extension as such, being able to move it self, or act upon it self.

[2] Ibid. 47. ...since no body could ever move it self; it follows undeniably, that there must be something else in the world besides body, or else there could never have been any motion in it.

Ibid. 844. All the local motion that is in the world, was first caused by some cogitative or thinking being, which not acted upon by any thing without it, nor at all locally moved, but only mentally...

[3] T. I. S. 667. (Generation and mundane system) requires a certain proportionate commixture of motion and rest; no sun, nor moon, nor earth, nor bodies of animals; since there could be, no coherent consistency of any thing, when all things fluttered and were in continual separation and divulsion from one another.

[4] (This ambiguity in the question of the infinite divisibility of matter may have its reason in that, although theoretically we are bound to assume infinite divisibility, yet in practise it can never be achieved. If it could be done, we could never attain to the smallest particles infinitely. We must admit that the sensible world is simply not able fully to realise reason and understanding.)

This is but one of many instances of the existence of two strands, a critical and an uncritical, which run through Cudworth's whole work. To the way in which these two strands are found, at one time running side by side, at another closely intertwined, is due much of the misunderstanding from which Cudworth suffered. He thereby made his philosophy unduly difficult of access.

A further property of the corporeal substance is solidity, by which, without any activating force, it prevents other bodies from penetrating it, thus preserving the absolute aliud extra aliud[1].

Although Cudworth was attracted to the atomic theory, he refrains from putting it forward dogmatically. He sees in it the only possible way of making the sensible world at all intelligible, and of arriving at any scientific knowledge of it[2]. Its main significance, however, is to prepare the way for a theistic explanation of the world by revealing how indigent and impotent the material world is. This impotence testifies to the existence of an incorporeal autokinetic substance, since life and understanding cannot reasonably be derived from dead matter.

Cudworth points out the profound inconsistency of an atomism based on atheism, which is a mere travesty of the genuine Pythagorean atomism first introduced by Democritus, Leukippus, Protagoras and Epicurus in ancient times and reintroduced in a modern form in the philosophy of Thomas Hobbes.

Atomism is an explanatory hypothesis of the sensible world which springs solely from reason[3]. Our sense experience would rather tell against it. It can, therefore, never be a popular way of thinking, and will be a help only to those who seek an explanation of the world by thought. Whoever rejects it will be forced to renounce all scientific knowledge of matter, but he will not thereby escape the awkward

[1] Ibid. 829. ... it is an essential property thereof to be antitypous and impenetrable, that is, to justle or shoulder out, all other extended substance from penetrating into it, and coexisting with it, so as to possess and take up the same room or space.

[2] Mor. 64. Wherefore unless we will assert that these lower sensible things are utterly incomprehensible to our understanding, whilst it is able clearly to comprehend things of a higher nature, we must conclude this old atomical... philosophy to be true.

[3] T. I. S. 27. This atomical physiology seems to have had its rise and origin from the strength of reason exerting its own inward active power and vigour, and thereby bearing it self up against the prejudices of sense, and at length prevailing over them.

consequence of acknowledging that the mind can understand the higher things, while the lower are beyond its grasp. We shall see later how much Cudworth himself tends towards this seemingly paradoxical standpoint.

I should like to draw attention here to two points in which Cudworth diverges from Plato, and especially from Plato's Timaeus to which he so often refers. We shall gain thereby an idea of Cudworth's critical attitude towards his own theory of substances.

1. As we have seen above, Cudworth, like Descartes, holds the view that we are justified in concluding the existence of bodies "without" the mind from the clear and distinct intelligibility of Extension and its modifications. This conclusion is considered a conclusion according to reason, which conveys absolute certainty.

From this it follows, firstly, that the material world is altogether intelligible; there is no sub-logical remnant which evades the grasp of reason. Secondly, that we cannot arrive at a direct and certain knowledge of the existence of the material world through the experience of the senses, but that this knowledge can, nevertheless, be attained by thought.

Cudworth agrees with Plato that by sensible experience we do not attain to any direct knowledge of the outside world[1], but that we are able to conclude its existence only by thought. Plato, however, says that this conclusion is not a conclusion according to reason but rather a "bastard" conclusion *νόθος συλλογισμός*[2], to which we are forced by the fact that our knowledge is never complete or perfect, whenever we attempt to explain the outside world by means of ideas. We are therefore led to conclude its existence in an indirect way as the cause of the deficiency of our knowledge. The material world thus appears not as a constituent, but as a deficient, cause of knowledge. Plato defines this deficient cause as space *χώρα* which is in itself absolutely shapeless, an indifferent womb ready to receive any form[3]. Cudworth, as we have seen, considers space as the constituent of bodies which alone makes them intelligible to us. Plato, on the contrary, holds that

[1] Mor. 72. ... sense alone is not the criterion or judge of what does really and absolutely exist without us.

[2] Plato Tim. 52*b*, cf. T.I.S. 48.

[3] Plato Tim. 50*d–e*.

space obscures them for us by pulling them down into a sub-logical sphere where reason cannot find them[1].

2. The decisive deviation from Plato is connected with this different conception of space. For Cudworth, the Becoming of the sensible world is due to the change of the accidents of the one indestructible substance, which is in no way itself touched by this Becoming, but remains unalterably the same. With such a hypothesis Cudworth is indeed very far from Plato, for whom space (χώρα) is not an unalterable substance with independent existence in itself, but on the contrary, the very element of Becoming in things[2]. Plato never assumed any substantial Being outside of idea, and it is interesting to see that Cudworth, according to his central principles of knowledge, does not do so either. But thereby he actually abandons his theory of substances which elsewhere he had so emphatically put forward.

This leads us to another remarkable difference between the two thinkers concerning the problem of immortality. For Cudworth, both the material world and the soul are, as substances, in themselves indestructible. Yet because they are created they depend entirely on the goodness of their Creator, who guarantees that what he has created, he will not destroy.

Plato does not admit that any Becoming has in itself stability or Being except that which, from moment to moment, it receives from the ideas[3]. All creation is wholly dependent upon the goodness of the creator, because it has no guarantee of subsistence in itself[4].

[1] Plato Tim. 49*e*.

(Space as such cannot be thought. It is not an EIDOS. Yet without space we cannot think, either corporeal things or even ideas; for even ideas we think as being one beside the other in a certain sense. As far as Cudworth follows Descartes, he interprets this necessary connection of corporeal things with space as to assume space to be the substantial constituent of the sensible things, by which alone they can become intelligible to us. It is difficult to imagine a more acute conflict between Cartesianism and Platonism in Cudworth. Later on we shall see how this same problem presents itself in the light of Cudworth's Platonism.

Cudworth, moreover, does not strictly think out Plato's critical examination of the knowledge of ideas in us, where it clearly appears that we can never quite free ourselves from space. We shall examine this question at length later on.)

[2] Plato Tim. 49*a*, 52*b*.

[3] Plato Tim. 50*d*; Phil. 26*d*.

[4] Plato Tim. 41*a–b*.

If we critically think out the atomic theory, we see at once, how difficult it is to hold to it. What can the substance Matter be, which is supposed to underlie its modifications? It is certainly never directly intelligible. Is it identical with atoms before they clothe themselves with their essential properties? What else can those naked atoms be but formless space; so formless that nothing at all can be said of them. How then can we predicate anything about them at all? It will become clear later on that with the theory of substances, and with atomism, we are by no means in the centre, but only on the outskirts of Cudworth's philosophy.

The Incorporeal Substance

The incorporeal substance as created Being depends directly on its creator[1]; not, however, as does matter, merely through the fact of its creation and preservation, but also by means of the direct participation in the self-activity of God, who himself is the source of life and understanding[2]. Because it is incorporeal, it is also indivisible, and therefore free from all the limitations which are characteristic of Extension. It is essentially energy and self-activity, yet not of itself, but by participation.

It is therefore capable of:

1. *Acting upon itself.* It can send itself out into multiplicity and recall itself into unity. It is unity in multiplicity, *πέρας* (limit) to every *ἄπειρον* (indefinite), giving it thereby meaning and purpose. Its activity can be symbolised by a circle, as against the straight-line progression of matter into purposeless infinity. The capacity of acting upon itself springs from its indivisibility. Cudworth, like Henry More, calls this action upon itself a reduplication of the soul upon itself[3],

[1] T.I.S. 831 ff.

[2] Ibid. 733. Of which one perfect mind and knowledge, all other imperfect minds (being derived from it) have a certain participation; whereby they are enabled to frame intelligible ideas, not only of whatsoever doth actually exist, but also of such things, as never were, nor will be, but are only possible; or objects of divine power.

[3] Fr. w. 71. We are certain by inward sense that we can reflect upon ourselves and consider ourselves, which is a reduplication of life in a higher degree.

which manifests itself in thought as self-consciousness, and in will as self-determination[1].

2. *Acting upon matter*. Matter in itself is dead, for it cannot produce motion. It is the incorporeal substance which is the source and directing force of local motion in matter, it determines its direction and speed, and is thereby principle of life in general[2].

3. The incorporeal substance is also the agent of thought. It is the source of thinking, filled with the restless desire of intending new objects and of awakening ever new ideas in itself. This desire is as proper to it as life itself[3]. The incorporeal substance as thought implies that which is actually known as well as that which is virtually knowable but yet unknown. It does not, however, contain anything which is in itself unintelligible and contradictory, for this would be a non-entity. Thus the incorporeal finite substance, either actually or virtually, embraces all Being[4].

It is characteristic of a finite mind that it can only lay hold on truth step by step and not without the stimulation of sense-experience. There are, therefore, in the soul many things only virtually knowable of which it cannot yet have become conscious[5].

The soul has, moreoever, vital energies which act cogitatively but unconsciously[6].

4. The incorporeal substance is also the principle of action[7], manifested as the self-determining power in every contingent and particular choice. It makes it possible for us to make a decision in a contingent alternative, to act before doubts are solved, and even against a clear and distinct judgment of reason.

[1] Ibid. 37. (This HEGEMONIKON) ... has a self-forming and self-framing power, by which every man is self-made into what he is, and accordingly deserves either praise or dispraise, reward or punishment.

[2] T. I. S. 864. It being also natural to souls as such, to actuate and enliven some body, or to be, as it were clothed therewith...

[3] Ibid. 846.

[4] Mor. 134. ... the soul is all things intellectually.

[5] Ibid. 135.

[6] T. I. S. 159.

[7] Ibid. 851.

The incorporeal substance is represented by the particular souls. Each soul is a centre of energy; a substance in itself[1], and therefore in itself indestructible. We have a direct consciousness of it in our 'inward sense', moreover the philosophical axiom that 'nothing can come from nothing', leads us conclusively to assume it[2]. Life, thought, and will, which we actually find within ourselves would be without a cause, if there were no incorporeal substance, since they cannot possibly be derived from dead matter.

We conclude the incorporeal nature of the soul from the manner of its action in sense perception and thought.

1. From sense perception[3]. We should get into great difficulties, if we tried to explain sense perception without admitting an indivisible centre in man, where all the perceptions converge. Were the soul Extension, it would be necessarily divisible. Each part would then be Soul itself, and the other parts superfluous. We should then be faced with three different explanations of sense perception, all leading to a paradox.

a) Either every part of the soul perceives one part of the object; but then we never attain to the perception of the whole.

b) Every part of the soul perceives the whole of the object, and, if the soul is infinitely divisible, the result is an indefinite number of perceptions of one and the same object. Yet experience shows that we have, from one particular point of view, only one single perception of one object.

c) The soul is considered as either a mathematical or a physical minimum, in which the whole object converges. The soul as a physical minimum, however, could not have distinct perceptions of different parts of the universe at one and the same time. If a physical minimum, then the soul is necessarily matter, and shares all its properties. That means that it is without life and without senses, and the paradoxical situation arises that, as Soul, it is simultaneously endowed with

[1] Ibid. 830. A thinker, is a monade, or one single substance, not a heap of substances...

[2] Ibid. 637. Whereas we are certain of the existence of our own souls, partly from an inward consciousness of our own cogitations, and partly from that principle of reason, that "nothing can not act".

[3] Ibid. 824–827.

reason and understanding. We may well ask, from where could such a soul-atom have come into man? and how could it preserve itself in the flux of the accidental changes of matter?

If Soul is divisible, and therefore a plurality, it follows that every man is a plurality of percipients. Indeed only as an incorporeal substance can the whole soul perceive the whole object; and only then can it be united with and act upon all the parts of the body at the same time.

From the nature of perception we may therefore infer in man and in animals an unextended centre in which all the perceptions converge; an indivisible centre which is always identical with itself, an 'ego' which embraces also the exterior part of man, and which is a self-active living power, an "inner entity" which holds everything together.

2. The incorporeal nature of the soul also follows from thought[1].

If Soul were Extension, it could only think Extension and would be incapable of conceiving larger and smaller things in the same manner. But this is contradicted by the fact, that we can form conceptions of unimaginable unextended and indivisible things, which can only be comprehended in a sphere of indivisibility; for instance, the geometrical conceptions of pure breadth as indivisible as regards thickness, of pure length as indivisible as regards breadth and thickness, and the mathematical point as indivisible in all respects.

Furthermore, we can form conceptions of intending powers and of moral ideas. The ideas or 'Natures' of things are indivisible. Even Extension is comprehended by the soul in an unextended way, because the soul can think only in ideas. So the thought of a long distance does not take more room in the soul than that of a mathematical point. Thoughts are neither measurable nor divisible. The soul is without magnitude, and therefore originally without any relation to place. It can nevertheless attain to such a relationship through vital union with a body, which is, however, always accidental to it. Only as an indivisible substance can it act upon different parts of the body in such a way that its own unity and identity are never affected.

[1] Ibid. 827–829.

Soul, as substance, is indestructible[1]. As created Being it has necessarily an ἀρχή (first cause), which must at least mean that it is dependent on a higher principle, if not also ,that it has a beginning in time.

Demonstrations of immortality can only show that the soul is a substance. Of its actual immortality no logical assurance can be attained, because in its existence it is entirely dependent on the will of God, and in his unchanging goodness lies its only hope of immortality[2].

Cudworth mentions three different opinions about the creation of particular souls; *a)* the soul created from a pre-existent soul substance; *b)* every particular soul existing from eternity as a particular soul, but entirely dependent on God; *c)* God creating every soul in a new creative act at the moment of the conception of the body.

The last hypothesis presents a difficulty in that it makes the creative act of God, as it were, dependent on the human act of procreation, and thus gives the natural generation a certain temporal priority over the spiritual. In spite of this Cudworth deems it the most reasonable hypothesis because it shows that the creative work of God has not

[1] Ibid. 848. (The soul according to Cudworth is immortal in its own nature, "since no substance of itself ever vanishes into nothing". That Cudworth can so trustfully consider the soul apart from the vital union as being beyond the process of genesis again lies in that he did not pay sufficient attention to how the ideas are activated in the soul. In his hard-fought battle against materialism he naturally puts all the emphasis on the immutability of the idea in itself.)

[2] Ibid. 45. ... we must then needs acknowledge, that all created being whatsoever, owes the continuation and perpetuity of its existence, not to any necessity of nature without God, and independently upon him, but to the divine will only. And therefore though we had never so much rational and philosophical assurance, that our souls are immaterial substances, ... yet we could not, for all that, have any absolute certainty of their post-eternity, any otherwise than as it may be derived to us, from the immutability and perfection of the divine nature and will, which does always that which is best. For the essential goodness and wisdom of the deity is the only stability of all things.

Ibid. 869. Wherefore the assurance that we have of our own souls' immortality, must depend upon something else, besides their substantiality, namely a faith also in the divine goodness, that he will conserve in being or not annihilate, all such substance created by him; whose permanent subsistence is neither inconsistent with his own attributes, nor the good of the universe; as this of rational souls unquestionably is not.

come to an end, and that he has not, as Descartes thought, framed the universe by one single creative act, to leave it afterwards to itself, while he merely looks on, while it runs its course as a perfect machine[1].

Extension and Cogitation are, according to Cudworth, not themselves substances, but merely attributes of the corporeal and incorporeal substances[2]. The corporeal substance is, however, identical with its attribute extension, and the two terms are interchangeable[3]. But a different relationship exists between the incorporeal substance and Cogitation.

Descartes assumed that each soul is a manifestation of the RES COGITANS, which embraces the whole range of consciousness, whereby self-consciousness is essential to it.

Cudworth, on the contrary, conceives the soul as a sphere of higher and lower functions, "powers" of the soul. They are determined and controlled by a focus of energy within the soul, the *ἡγημονικόν* or *αὐτεξούσιον*. In this focus the soul is a unity in multiplicity, perfectly indivisible, perfect simplicity. It is not itself these powers, but it wields them and rules over them. Furthermore, by intention, it determines also its passive qualities[4]. The hypothesis of the *αὐτεξούσιον* explains, moreover, the different reactions of men to the same judgement of reason. It also gives a splendid foundation for free will and, in consequence, for the capacity of the proposing of purposes and of the possibility of theoretical and practical mistakes, error and sin[5].

[1] T. I. S. 43–44.

[2] Ibid. 47. Wherefore it must be needs granted that life and cogitation are the attributes of another substance distinct from body and incorporeal.

[3] Ibid. 770.... space... the very extension of body it self.

[4] Fr. W. 35–36. Moreover we know, by certain experience, that speculation and deliberation about particular things is determined by ourselves both as to objects and exercise. We can call it off from one thing, and employ it or set it a work upon another, and we can surcease, suspend and stop the exercise of it when we please too, diverting ourselves into action. From whence it is plain that there is something in us superior thereunto, something more universal and comprehensive, and yet withal more simple, which is hegemonic to it and does manage and determine the same.

[5] Muirhead, The Platonic Tradition, extracts from Cudworth's manuscripts, 64–65.

...there are in the soul lower and higher principles of life and action, appetites, passions... there must of necessity be in the soul one common focus or centre in which

Only man, endowed with free will, can be the object of divine justice, worthy of praise or blame, without which there could be neither morality nor religion.

The *αὐτεξούσιον* marks one of the central points in Cudworth's philosophy where Logic, Ethics, and Religion converge and where the finite part of their metaphysical foundation lies.

The *αὐτεξούσιον* is a centre of energy rather than a fact. As such it cannot become a direct object of knowledge, because it necessarily precedes every act of thinking[1]. It can, however, indirectly be inferred, for, if we were without it, we should either be mechanisms or pure cogitation. As mechanisms we should be determined from outside only; as pure cogitation we should be unceasingly occupied in pure thought, ever forming necessary connections of ideas in a linear progression. We should then be incapable of any communication with the outside world, and so incapable of acting at all, and deprived of presence of mind, without the faculty of intention and without memory, since the recalling of past ideas is made impossible by the necessary progression of thought. Both these statements are contradicted by experience.

In spite of his careful exposition of the difference between cogitation and the incorporeal substance, Cudworth does not keep strictly to it in the use of his terms.

Plastic Nature

The problem of interaction between the corporeal and the incorporeal substance is the crucial point in the Cartesian dualism of substances. Descartes tried in vain to solve it by assuming some obscure mediating functions of the "esprits animaux" in the Conarium. This unsolved problem has again and again stimulated the followers of Descartes, and Cudworth also was much preoccupied by it.

all these kinds may meet, some one thing in which all is recollected and knit together, something which is conscious of all congruities, both higher and lower, of all cogitations, powers and faculties of the soul... which also can wield, steer and guide the whole soul and exercise power and dominion over the several faculties of it.

Now this is the whole soul redoubled on to itself, which both comprehends itself and, holding itself as it were in its own hands, turns itself this way and that way.

[1] Plato. Theait. 196*e* (about the necessary defectiveness of method).

It is clear that no direct transition from one substance into the other can take place. Cudworth, nevertheless, assumed a reciprocal influence of one substance upon the other in the intermediary sphere of the lower energies of soul. He believes that they can unite themselves to the body in a vital union, and suffer all that the body suffers, yet have no share in the pure passivity of matter, neither are they anchored in the pure activity of reason.

Cudworth says that these lower powers of the soul must not be altogether separated from its higher functions, as did Descartes in his attempt to expound the sensitive soul as a mechanism[1]. For if any life is held to be automatically produced from matter and local motion, there is not sufficient reason to stop at deriving will and reason also from the same source. The soul is the source of life in the same degree in which it is the source of thought and will. This has its foundation right in the centre of Cudworth's philosophy in the theory that intelligibility of idea is Being, and Being is intelligibility. We shall recur to this later on.

Cudworth sees the decisive mistake of Descartes in that he reduced the substance Cogitation to self-consciousness. Therein lay the ultimate cause of his considering animals as mere machines. As far as he could plausibly assume that animals have no self-reflection he actually could conceive them as mere automatons.

The soul can unite itself to its body in an intimate vital union. By this union, imperfect created beings are given a greater perfection, since by means of it they are made capable of communing with the outside world through the senses and imagination[2].

Soul, as self-activity and source of life and thought, has substantial life in itself, which remains untouched by the process of generation[3].

[1] T. I. S. 44. ...for either all conscious and cogitative beings are incorporeal, or else nothing can be proved to be incorporeal.

[2] Ibid. 818. ...the imperfection of whose nature necessarily requires the conjunction of some body with them, to make them up complete; without which it is unconceivable, how they should either have sense or imagination.

[3] T. I. S. 842. Ibid. 862. We speak not here of that life (im-properly so called) which is in vulgar speach attributed to the bodies of men and animals: for it is plainly accidental to a body to be vitally united to a soul or not.

Therefore is this life of the compound, corruptible and destroyable, without the destruction of any real entity; there being nothing destroyed, nor lost to the universe,

Since it is incorporeal, it has no direct relationship to space and can exist apart from a body. Nor is there any reason for doubting that it can inhabit and quicken also bodies of different structures from ours. Only when united to a material body does it become bound and limited to a local field of action, and acquires an accidental life, which however, it loses again when the material vehicle dies. This dying does not essentially affect it in any way, since the union with that corruptible body has always been but an accidental one[1].

It is essential and natural for a finite soul to be united to a body as its vehicle and field of action. An absolute separation of soul from body is not its ultimate end of perfection and beatitude[2].

Cudworth discusses the question of aetheric and aeric bodies[3]. He regards the dogmatisation of the middle ages of the absolute separation of body and soul in death as less sound[4] than the opinion that an aetheric body essentially belongs to each soul and that it is immortal like the soul itself. He believes that no created soul can ever be entirely deprived of such a vehicle, or else no sphere of action would be left to it, and it would then fall into a stupor of inactivity contrary to its own nature, and would moreover have to be forcibly held down in this state.

These are speculations familiar to 17th century minds. They are of less interest to us in our present research. Yet of great importance is the emphasis which Cudworth lays on the fact that a finite soul needs a finite and limited field of action, not however, because the body

in the deaths of men and animals, as such; but only a disunion or separation made, of those two substances, soul and body, one from another.

But we speak here of the original life of the soul it self that this is substantial, neither generable nor corruptible, but only creatable and annihilable by the Deity.

[1] Ibid. 830. For though that secondary and participated life... in the bodies of animals be indeed a mere accident, and such as may be present or absent without the destruction of its subject; yet can there be no reason given, why the primary and original life it self, should not be as well a substantial thing, as mere extension and magnitude.

[2] Ibid. 794. That the most consummate happiness, and highest perfection, that humane nature is capable of, consisteth not in a separate state of souls, strip'd naked from all body, and having no manner of commerce with matter; as some high-flown persons in all ages have been apt to conceit.

[3] Ibid. 785–822 (about different vehicles of the soul).

[4] Ibid. 815.

is its principium individuationis, but because only thereby can a finite corporeal substance fulfil the purpose for which it was created, which is to act as a finite transcendental cause [1]. The principium individuationis and the identity of human personality is not necessarily bound to the numerical identity of the material part. Each soul in itself is a proper eidetical unity [2].

The problem of the manner of interaction between body and soul, and of the intermediary links, can be examined either from the ontological or the gnoseological point of view. In the first case the emphasis is primarily on the soul as the life-giving principle of the body; in the second case on the soul as the principle of thought and of the knowledge of the outside world. Here we shall examine the question from the ontological aspect.

The problem does not concern only the relationship of particular souls with their bodies, but also the relationship between the creator and matter [3]. Cudworth does not admit any parallelism between MACRO- and MICROCOSMOS in the sense that the relationship of the creator, conceived as World Soul with his body the universe, corresponds with the relationship of particular souls with their bodies [4].

[1] Ibid. 794. ... being essentially but parts of the universe, and therefore not comprehensive of the whole; finite or particular, and not universal beings; ... the sphere of their activity, could not possibly extend to any further, than to the quickening and enlivening of some certain parts of matter and the world, alloted to them; and thereby of becoming particular animals; it being peculiar to the deity, or that incorporeal substance, which is infinite, to quicken and actuate all things.

Ibid. 798. But thus much is certain, that our humane souls were at first intended and designed by God Almighty, the maker of them, for other bodies and other regions; as their proper home and country and their eternal resting place.

[2] T. I. S. 798–9. ... individuation and sameness of men's persons does not necessarily depend, upon a numerical identity of all the parts of matter, because we never continue thus the same, our bodies always flowing like a river, and passing away by insensible transpiration...

[3] Ibid. 668–9. Wherefore in the bodies of animals, the true and proper cause of motion, or the determination thereof at least; is not the matter it self organized; but the soul either as cogitative or plastickly self-active, vitally united thereunto, and naturally ruling over it. But in the whole world it is either God himself, originally impressing a certain quantity of motion upon the matter of the universe, and constantly conserving the same; ... or else it is instrumentally, and inferiour created spirit, soul or life of nature, that is, a subordinate hylarchical principle, which has a power of moving matter regularly, according to the direction of a superiour perfect mind. [4] Ibid. 841 (concerning the refutation of God as the World Soul).

When God creates new Being he always acts as ψυχὴ ὑπερκόσμιος and never as ψυχὴ ἐγκόσμιος. He does not immerse himself in his creation, nor is he bound to it like a created soul to its body. A certain parallelism cannot be denied in so far as the finite soul participates in the creative power of God and realises this divine power in a finite way according to its capacity[1].

Yet a vital intermediary sphere of Plastic Nature does not only exist between body and soul, but also between God and matter[2]. Cudworth uses many different terms for it; all of them are to be taken rather as images than as definitions, and they should not therefore be pressed. They are merely meant to show various aspects in a tentative and complementary way.

Plastic Nature is the lowest energy of the soul able to enter into a vital union with matter. It then rules matter by the laws of reason from within[3]. Laws of nature are not forced upon matter from outside by an almighty will which is intrinsically alien to it[4], but they are natural and proper to matter and perfectly correspond to its nature by the strength of this vital union.

[1] Ibid. 747. But since it is certain, that imperfect created beings can themselves produce some things out of nothing preexisting, as new cogitations, and new local motion, new modifications and transformations of things corporeal, it is very reasonable to think, that an absolutely perfect being could do some thing more; that is, create new substances out of nothing, or give them their whole being.

("Created out of nothing" means for Cudworth created out of God alone.)

[2] Ibid. 147–74 (Digression about Plastic Nature).

Ibid. 172–4 (Summary of the digression).

[3] Ibid. 155 *b*. Knowledge and understanding, in its own nature is a certain separate and abstract thing, and of so subtil and refined a nature, as that it is not capable of being incorporated with matter, or mingled and blended with it.

But nature is RATIO MERSA ET CONFUSA, reason immersed and plunged into matter, and as it were fuddled in it and confounded with it.

[4] Ibid. 162. (Plastic Nature: the true and proper fate of matter.)

Ibid. 207. (Things in the universe) they being all naturally subject to this (God's) authority, and readily obeying his laws.

("Law of nature" is, according to Cudworth, divine law.)

Ibid. 147. Though it be true that the works of nature are dispensed by a divine law and command, yet this is not to be understood in a vulgar sence, as if they were all effected by the mere force of a verbal law or outward command, because inanimate things are not commandable or governable by such a law. Cf. Plato. Tim. 48 *a*.

... there must needs be some other immediate agent and executioner provided...

Plastic Nature is unconscious of itself, and acts by the continuous inspiration of divine reason towards purposes determined by God[1]. It is comparable to a craftsman who carries out the design of an architect; to an instrument in the hands of an artist; or, again, to the artist himself, although it works far more perfectly than any human artist, for it realises its ends unhesitatingly and with assured success. There is no struggling with its material, for that is not alien to it. On the other hand, it does not reach to the highest in human art, because it works unconsciously and without setting its own purposes.

Plastic Nature works in direct dependence upon the divine mind. It is directed, and often corrected, by God; for, being the lowest energy of the soul, it sometimes makes mistakes. At times it is altogether over-ruled by God in a most marvellous way.

Plastic Nature explains why nature works imperfectly. Being, as it is, a created cause, it is not almighty. The limitations of its creaturely powers are to be seen, for instance, in the slowness of all natural processes. If these were brought about by an omnipotent cause, which could just as easily create everything in one explosive act, the slowness, Cudworth says, would be a somewhat pompous, affected and unnecessary show[2]. The necessity of Plastic Nature is evident also from another consideration; created Being, if it were exposed to an almighty creative power, would necessarily be crushed by it. The almighty power must therefore restrain itself and work through a created efficient cause. Plastic Nature thus reveals the indispensable self-limitation of God in his act of creation and preservation of the world.

Considered as a craftsman, Plastic Nature gives an answer to the startling objection of Thomas Hobbes, when he says that Providence, if there were such a thing, could never enjoy the beatitude of its own

[1] Ibid. 150. The wisdom of God will not be shut up nor concluded wholly within his own breast, ... but will print its stamps and signatures every where throughout the world.

Ibid. 680. ... neither are all things performed immediately and miraculously by God himself, neither are they all done fortuitously and temerariously, but regularly and methodically for the sake of ends, though not understood by nature it self, but by that higher mind which is the cause of it, and doth as it were continually inspire it.

[2] Ibid. 150.

perfection, because it would be continually harrassed by the arranging of every detail in the order of the universe [1].

Cudworth, of course, does not take such crude arguments too seriously. It is therefore unnecessary to enlarge on them. The philosophical significance of Plastic Nature lies in the fact that there is no automatic mechanism in nature which does not move towards its final cause in direct dependence upon God [2]. In other words there is no such thing as dead matter in the organism of the universe; no Being which does not subsist between reason as its terminus ad quo and reason as its terminus ad quem; no Being which does not bear the seal of the divine mind, whose secret melody, heard by the spirit alone, pervades the universe [3]. We shall find this argument to be of great importance in the refutation of Descartes.

[1] (The following quotations may show how little dogmatic weight Cudworth attaches to his descriptions of Plastic Nature.)

Ibid. 884. Nevertheless the minds of weak mortals may here be somewhat eased and helped by considering... that there is no necessity, God Almighty should do all things himself immediately and drudgingly; but he may have his inferior ministers and executioners under him, to discharge him of that supposed encumberment.

Ibid. 886. The activity of the Deity as a perfect being, is altogether as easy to it, as its essence.

On the other hand, ibid. 884 ... nor yet the Deity be supposed to do everything it self immediately and miraculously, without the subservient ministery of any natural causes; which would seem to us mortals, to be not only a violent, but also an operose cumbersome, and moliminous business.

[2] Ibid. 672. But still so, as that all is supervised by one understanding and intending cause, and nothing passes without his approbation; who when either those mechanick powers fall short, or the stubborn necessity of matter proves uncompliant, does overrule the same, and supply the defect thereof, by that which is vital; and that without setting his own hands immediately to every work too; there being a subservient minister under him, an artificial nature, which as an ARCHAEUS of the whole world, governs the fluctuating mechanism thereof, and does all things faithfully for ends and purposes, intended by its director.

[3] T. I. S. 151.

(Cudworth sees in the *συναίτια* of Plato a parallel to Plastic Nature. This may be correct in so far as Plastic Nature is considered as the inner fate of nature, *ἀνάγκη*; the comparison is inadequate in that Plastic Nature is contrasted with mechanics as a vital energy working for a purpose. This may show how complex Cudworth's descriptions of Plastic Nature are, at least when they are taken apart from the main lines of his philosophy and considered by themselves.)

In spite of all this long, and at first sight so plausible, exposition of the mediating part, played by Plastic Nature, Cudworth cannot give an answer to the decisive question, in what does the vital union between the lowest energy of soul and matter consist? He evasively calls it "a secret teaching of nature"[1].

The essential gap between matter and spirit can never be bridged by gradual transition, not even by the lowest power of the soul[2].

The true significance of Plastic Nature in Cudworth's philosophy is found less in its original function of interaction, than in the emphasis laid upon the final cause as the only true and reasonable cause[3].

By assuming Plastic Nature as the efficient cause of matter, Cudworth hoped to avoid two paradoxes from which Descartes did not escape[4]: that either natural laws execute themselves, or that God as the efficient cause "moves every stone". In order to evade the absurdity of a constant succession of miracles implied in the latter, Descartes arrived at his mechanical explanation of nature. He thereby misinterpreted the condicio sine qua non as a genuine cause; thus depriving the universe of its only true and reasonable cause, he abandoned it to mere chance.

We shall examine this question more thoroughly in connection with the refutation of Descartes' rejection of the cosmological demonstration of the existence of God. We shall then see Plastic Nature in yet another aspect, for Cudworth the most important one, that of its being the cause of the perfection and beauty, the *εὖ καὶ καλῶς* of the universe.

[1] Mor. 112.
(Causality is not intelligible for us, it works "by the secret instinct of nature".)
Cf. Plato Tim. 68 *d*.

[2] (Cudworth goes further and denies also a gradual transition from a simple cogitation to self-consciousness.)
T. I. S. 173. Neither can the duplication of corporeal organs be ever able to advance that simple and stupid life of nature into redoubled consciousness or self-perception; nor any triplication or indeed milleclupation of them, improve the same into reason and understanding.

[3] (J. H. Muirhead considers this treatise on Plastic Nature as a proof of the modern trend in Cudworth's thought. Modern biological research moves steadily towards the conviction that no life whatever can be explained mechanically, and that there is no such thing as dead matter at all. The Platonic Tradition, 38.)

[4] T. I. S. 147.

Chapter II

THE THEORY OF KNOWLEDGE

On analogy with the dualistic theory of substances, Cudworth also divides the knowledge of reality into two different parts: the passive perception of the senses, and the active perception of reason; between them lies the intermediary sphere of imagination, corresponding to Plastic Nature as the link between the two substances. Experience of the outside world reaches us through the mediating function of imagination and, together with sense, it leads to the activity of pure thinking where alone the perceived object is truly comprehended. Cudworth illustrates these three stages of knowledge by an example: let one and the same object be held up to a mirror, to a living eye, and to reason, and see what happens [1].

1. The first stage is *sense perception* (the mirror). It is pure passivity and achieves no more than the reflection of the object in the eye as in a mirror. In this process local motion which comes from the object is transferred to the nerves of the eye, and from them to the brain. In the brain the "esprits animaux" are stirred, and by their motion the magic transition of sense perception to the vitally sympathetic, passive-active imagination is achieved.

2. The second stage is *imagination* (the living eye). The living eye surpasses the passivity of the mirror in so far as it is conscious of what it perceives. Conscious perception or imagination is a knowledge actively reproduced by the soul of that which the body experiences. It is not explicable by mere passive transference of local motion, because the soul as incorporeal substance cannot receive local motion in a direct way [2]. All the impressions, as far as we are conscious of them, are, therefore, mediated in a twofold way.

[1] Mor. 150–6. Mor. 75–81.

[2] Ibid. 112. ... because by sense the soul doth not suffer immediately from the objects themselves, but only from its own body, by reason of that natural and vital

a) By the "esprits animaux" which are the seat of the vital union. They receive the local motion from the nerves, and by their motion they awaken the imaginative power. Cudworth obviously perceived the highly speculative nature of the hypothesis of the "esprits animaux" and admits that we hardly know anything of these secrets of vital union.

b) By the imaginative power. The second mediation is achieved by the active contribution of the imaginative power which, incited by the "esprits animaux", actively reproduces an imaginative picture in a "confused" manner.

This is however not achieved by a free act of thinking because the soul cannot choose the object which affects it. There exists, moreover, most probably a certain connection, determined by natural law, between local motion and impression, in the sense that a certain local motion calls forth the reproduction of a certain corresponding image in the soul[1].

This determination of imagination is, however, weakened by the fact that the soul directs also its passive powers through the function of the *αὐτεξούσιον*. In a free act of intending an object it determines in advance sense perception to a certain degree[2]. We might ask whether any conscious sensation would be possible at all without this preconceiving act of the soul. So much is certain, that without the activity of the imaginative power, we could never become conscious of any object perceived by passive sense-perception.

From this it follows that by sense perception we never attain to any direct knowledge of bodies and their motion; neither have we any

sympathy which it hath with it, neither doth it suffer from its own body in every part of it, or from the outward organs of sense immediately; as from the eye when we see, the tongue when we taste, or the exteriour parts of the body when we feel, but only in the brain, or from the motions of the spirits there.

[1] Mor. 85. ... there being a necessary and fatal connection between certain motions in some parts of the enlivened body, and certain affections or sympathies in the soul.

[2] Fr. W. 27. ... but we are sensible that our minds are employed and set at work by something else, that we apply them both on contemplation and deliberation to this or that object, and continue and call them off at pleasure, as much as we open and shut our eyes, and by moving our eyes determine our sight to this or that object of sight.

criterion, whether, or in what degree, our confused perception corresponds with the object outside.

A further difficulty arises from the fact that the motions of the "esprits animaux", by which the imaginative power is incited, are not necessarily caused by an accidental local motion from outside[1]. The "esprits animaux" can also be incited from inside by the imaginative power itself and, independently of sense perception, they then react upon this same imaginative power which had first incited them. This may happen involuntarily or voluntarily; voluntarily, for instance, when past sensations are remembered, or when images are awakened by thinking; involuntarily by irrational bodily functions, real or imaginary[2].

If this is true we have through sense perception no objective criterion of the outside world whatever[3], and the doors are thrown open to any error and illusion in regard to the inciting causes. We do not know whether we are awake or dreaming, or indulging in phantasies. Real and imaginary sensations often go side by side, differing only in strength and constancy. The soul spontaneously believes the strongest. In sleep, when incitements from outside cease, the imaginary pictures become clearer and more constant, and the soul therefore often mistakes them for real images.

Irrational tendencies in the body, or abnormal nervous dispositions may lead to an excessive growth of the imaginative energies of the soul. This can produce a total loss of discrimination between real and imaginary impressions, because the normal proportions of strength and faintness, of flux and constancy of the images, is destroyed. He who gives free rein to his irrational tendencies risks losing, not only knowledge of thought, but also knowledge of the sensible world. And in the end he will find himself sadly caught in the labyrinth of ever-changing flights of illusive impressions in his own soul.

[1] Mor. 112 ff.

[2] Ibid. 82–3.

[3] Ibid. 72. See p. 9, note 1.

T. I. S. 780. But that there is something unimaginable even in body itself is evident, whether you will suppose it to be infinitely divisible or not.

Plato. Theait. 158*c*.

We notice that Cudworth, on the one hand, refuses to attribute irrational tendencies to the soul, preferring to have them spring from an obscure kind of "witches' cauldron" in the body, while in his theory of the *αὐτεξούσιον* he follows Plato's conception of two tendencies in the soul, a rational tendency symbolised by a circle, and an irrational, which, as a straight line, loses itself into the indefinite (*ἄπειρον*) [1].

Cudworth believes that a sound and watchful soul is capable of discriminating between real and illusionary impressions, but nevertheless that no knowledge, either of the existence or nature of the outside world, can be obtained through the senses [2].

The only thing certain is the fact of perception as a variously mediated "passion" of the soul; of the content of this perception we can predicate nothing.

Our capacity of perception and imagination is, moreover, limited in more than one way; for instance, in the acuteness of the senses, where indeed we are often far surpassed by the animals; and also by the narrow realm over which it extends. We can perceive and imagine only a very small section of the sensible world which lies between the great and the small; but in thought we are free to overstep this limitation [3]. The decisive limitation of sense perception, however, is that in it subject and object always fall apart [4]. The perceived object always remains exterior and alien to the soul. Our perception, moreover, only reaches to the surface of the things, its realm is altogether narrower than that of reason [5]. We can, for instance, think a figure

[1] Muirhead Manuscr. 64. We may will the irrational.

Plato. Nom. 898 *a–b*.

[2] T. I. S. 719. It is true indeed that sense considered by it self, doth not reach to the absoluteness either of the natures, or of the existence of things without us, it being as such, nothing but seeming, appearance and phancy.

[3] Mor. 213. Nay, he (the astronomer) cannot for his life have a true phantasm of any such magnitude which contains the bigness of the earth so many times, nor indeed fancy the earth a hundredth part so big as it is.

[4] Ibid. 98. But sense is of that which is without, sense wholly gazes and gads abroad, and therefore doth not know and comprehend its object, because it is different from it.

[5] Ibid. 160.

with a thousand corners, but to imagine it is beyond our capacity [1]. There always remains in sense perception a dim and confused part, where the soul is passively affected, and over which it never attains complete mastery.

Its main function is to maintain the ordinary course of practical life [2]. Its contribution to knowledge is, that it presents problems to the mind, urging it thereby to awaken ideas for the explanation of the sensible world [3].

The question of Truth cannot as yet be raised in sense perception. Each sense perception is true in that it is a certain "passion" of the soul, but its content remains strictly relative to the percipient. Sense perception is, as Cudworth puts it, "flat in the object", indeed rather below it, and lacks all power of discernment [4]. But without a criterion of objectivity different percipients can never reach a consensus concerning the reality and nature of perceived objects. No such consensus can exist either between man and man, or between man and animal; indeed even the same percipient judges the same sense perceptions differently at different times. Sense perception thus remains alto-

[1] Ibid. 212. And so as we can in like manner clearly understand in our minds, a thing with a thousand corners, or one with ten thousand corners, though we cannot possibly have a distinct phantasm of either of them.

[2] T. I. S. 637. We grant indeed that evidence of particular bodies, existing hic et nunc, without us, doth necessarily depend upon the information of sense.

Plato. Phil. 62*b*.

[3] Mor. 92–3. Wherefore though sense be adequate and sufficient for that end which nature hath designed it to, viz., to give advertisement of corporeal things existing without us, and their motions for the use and concernment of the body, and such general intimations of the modes of them, as may give the understanding sufficient hints by its own sagacity to find out their natures, and invent intelligible hypotheses to solve those appearances by.

Ibid. 94. Sense is but the offering and presenting of some object to the mind, to give it an occasion to exercise its own inward activity upon. Which two things being many times nearly conjoyned together in time, though they be very different in nature from one another, yet they are vulgarly mistaken for one and the same thing, as if it were all nothing but meer sensation or passion from the body.

[4] Mor. 94–5. But sense which lies flat and grovelling in the individuals, and is stupidly fixed in the material form, is not able to rise up or ascend to an abstract universal notion; for which cause it never affirms or denies any thing of its object, because in all affirmation, and negation at least, the predicate is always universal.

gether unsatisfactory, and urges to further research beyond its own scope[1].

3. *Reason.* The object which the mirror passively reflected, and the eye consciously perceived, reason comprehends again, not passively nor sympathetically, but in pure activity; turning away from the sensible object, it seeks the explanation of the object by reflecting upon itself.

As a further explanation of these three steps of knowledge, Cudworth gives an analysis of the three different acts of apprehension of a white triangular figure[2].

a) Sense perception sees triangular and white in a confused way as one, and that is the whole contribution it can make.

b) Imagination, assisted by reason, achieves the analysis of the compound object by distinguishing the attributes: white and triangular, as well as the corporeal substance to which they belong.

c) Reason, turning away from the particular white triangular figure, comprehends it by subsuming it under universal ideas.

The corporeal substance it comprehends as extension.

The whiteness it comprehends as a confused perception of the soul, caused by a certain constellation of atoms. It thus reduces the confused perception to the distinct ideas of the modifications of extension.

By definition reason further comprehends the essence of triangle in general, and of this given triangle in particular, according to its species. Thus it discerns as its general essence the pure geometrical plane.

Lastly, reason determines the relation of this particular triangle to other triangles, and in addition the proportions existing within this particular triangle.

Every imagined triangle is of a particular kind. It is therefore impossible to give a universal definition of it. Reason, however, has the capacity to form the universal idea of triangle, which is independent

[1] T. I. S. 634–5. Whereas the mind of man remaineth altogether unsatisfied, concerning the nature of these corporeal things, even after the strongest sensations of them, and is but thereby awakened, to a further philosophick enquiry and search about them, what this light really should be.

[2] Mor. 193. f.

of imagination or any personal idiosyncrasies of the thinker, and is precisely the same at all times in every thinking mind.

The only realm of Truth is the realm of universal ideas and their inter-relations. It can therefore only be attained by thought; since Being, intelligibility, and perfection grow and diminish in strict proportion, it can be said that reason has more reality than sense perception and a far higher perfection[1].

The Ideas

Cudworth distinguishes three kinds of ideas[2].

Ideas whose SINGULARIA fall under sense perception. Cudworth calls these the "sensible ideas".

Ideas whose SINGULARIA are neither originally, nor necessarily sensible, as for instance geometrical figures.

Ideas which cannot be represented by a sensible form, as for instance conceptions of inter-relations or energies of Soul; also sense perception itself, life, thought and will.

Cudworth adopts this distinction of ideas from Descartes, but he does not, as Descartes, derive the sensible ideas from sense perception. He deems it a futile work for reason to extract ideas from sensible things, thus, as it were, hammering out the sensible world into thin little leaves. If, by so doing, reason does not know what it is doing, it must be considered a bad craftsman, but, if it knows what it is doing, it must have preconceived the idea previously, and so its labour of extraction is altogether superfluous[3].

[1] Ibid. 296. ... the mind and intellect is in it self a more real and substantial thing, and fuller of entity than matter and body.

[2] T. I. S. 732.

[3] Mor. 220 ff.

(The ideas are not abstracted from the sensible world.)

(It is an error) that then they suppose the intelligible ideas... to be made out of these sensible ideas and phantasms thus impressed from without in a corporeal manner likewise by abstraction or separation of the individuating circumstances, as it were by the hewing off certain chips from them, or by hammering, beating or anvelling of them out into thin intelligible ideas; as if solid and massy gold should be beaten out into thin leaf-gold.

To which purpose they have ingeniously contrived and set up an active understanding, like a smith or carpenter, with his shop or forge in the brain, furnished with

Cudworth considers the third kind of the greatest importance, that of the pure ideas. By means of them he proves in various ways that ideas do not have their origin in the sensible world, and that they cannot be engraved on the soul from outside, as on a tabula rasa[1], and that therefore it is impossible for reason either, to be derived from dead matter.

Cudworth does not attempt a systematic account of all pure ideas, for this would contradict the principles of his philosophy, that the activation of discursive reason is fundamentally dependent on the experience of the senses, in the sense, that the problems, presented by sense perception, incite reason to awaken ideas within itself[2]. And, since our experience is never completed, we do not know either, whether in the future new pure ideas will not be awakened[3].

By participation in the divine NOUS reason has a virtual omniformity to all Being. Idea is Being; consequently ideas are the Being, the essences, of all things. Virtual omniformity means, that reason has the capacity of activating in itself all the ideas; not, however, all at once, as befits a perfect mind, but one after the other. It can do this, because all the ideas are actually and eternally thought by the divine NOUS in one pure act of thinking.

all necessary tools and instruments for such a work. Where I would only demand of these philosophers, whether this their so expert smith or architect, the active understanding, when he goes about his work, doth know what he is to do with these phantasms before-hand, what he is to make of them, and unto what shape to bring them? If he do not, he must needs be a bungling workman; but if he do, he is prevented in his design and undertaking, his work being done already to his hand; for he must needs have the intelligible idea of that which he knows or understands already within himself; and therefore now to what purpose should he use his tools, and go about to hew and hammer and anvil out these phantasms into thin and subtle intelligible ideas, meerly to make that which he hath already, and which was native and domestick to him?

[1] Mor. 148–9.

[2] Muirhead Manuscr. 64. ... in perception... we employ imagination which is... an autokinecy..., occasioned and invited, but not caused from something without.

[3] (The impossibility of giving a final account of the pure conceptions of reason follows immediately from the principles of Cudworth's philosophy. With this question, however, he never dealt explicitly.)

We can never, therefore, create new Being in thought[1]. Cudworth once said, that even the divine mind cannot transcend its own content of ideas[2]. This, of course, is a paradox. How could there be anything added to the infinite and perfect? Cudworth probably meant it as an attempt to express in negative terms the positive infinity and perfection of the divine NOUS, which can never adequately be put into words.

Ideas are not innate entities of the mind, comparable to a sort of storehouse of ideas at man's free command, as Descartes thought. Cudworth has repeatedly been misunderstood here. In a certain way, the ideas are present to our mind in its virtual power of awakening in itself all ideas. Cudworth calls this the native cognoscitive power of the mind[3]. In this sense one can assume in each act of knowledge an

[1] T. I. S. 694. For the mind cannot make any new cogitation, which was not before, but only compound that which is.

And here we deny not, but that the humane soul has a power of compounding ideas and things, together, which exist severally, and apart, in nature, but never were, nor will be, in that conjugation: and this indeed is all the feigning power that it has.

In like manner that more subtle painter and limner, the mind and imagination of man, can frame compounded ideas and things, which no where exist, but yet his simple colours notwithstanding, must be real.

(The assumption that the imaginative power can join together ideas and things presents indeed a great difficulty. It contradicts Cudworth's central supposition, namely, that we can apprehend also the sensible things only by ideas; that we can therefore only connect ideas together.)

[2] Ibid. 695. Nay we conceive that a theist may presume with reverence to say, that God Almighty himself, though he can create more or fewer really existent things, as he pleaseth, and could make a whole world out of nothing, yet can he not make more cogitation or conception, than is; or was before contained in his own infinite mind and eternal wisdom; nor have a positive idea of any thing, which has neither actual nor possible entity.

[3] T. I. S. 733. (See p. 11, note 2.)

Mor. 134 ff. So all created intellects being certain ectypal models, or derivative compendiums of the same; although they have not the actual ideas of all things, much less are the images or sculptures of all the several species of existent things fixed and engraven in a dead manner upon them; yet they have them all virtually and potentially comprehended in that one cognoscitive power of the soul, which is a potential omniformity, whereby it is enabled as occasion serves and outward objects invite, gradually and successfully to unfold and display it self in a vital manner, by framing

act of remembrance, *ἀνάμνησις* as a necessary presupposition to all teaching and learning[1]. It is not an *ἀνάμνησις* of ideas, which have actually been thought at a certain time in life, or in any pre-existence, from whence they had fled into mere virtuality[2], but an *ἀνάμνησις* of ideas, which slumber in the cognoscitive powers of the soul and are then awakened and brought into consciousness. This can only happen by virtue of the participation of the finite mind in the infinite creative power of the divine mind, and in the ideas, which there are eternally

intelligible ideas or conceptions within it self of whatsoever hath any entity or cogitability.

Ibid, 126. ... knowledge is an inward and active energy of the mind it self, and the displaying of its own innate vigour from within whereby it doth conquer, master and command its objects, and so begets a clear serene, victorious, and satisfactory sense within it self.

[1] Ibid. 129. And this is the only true allowable sense of that old assertion that knowledge is reminiscence, not that it is remembrance of something which the soul had some time before actually known in a preexistent state; but because it is the mind's comprehending of things by some inward anticipations of its own, something native and domestick to it, or something actively exerted from within it self.

T. I. S. 693. But this argues of their great ignorance in philosophy to think that any notion or idea, is put into men's minds from without, meerly by telling, or by words; we being passive to nothing else from words, but their sounds and the phantasms thereof; they only occasioning the soul to excite such notions, as it had before within it self (whether innate or adventitious) which those words by the compact and agreement of men were to be signs of; or else to reflect also further upon those ideas of their own, consider them more distinctly, and compare them with one another.

And though all learning be not the remembrance of what the soul once before actually understood, in a preexistent state... yet is all human teaching, but maieutical, or obstetricious; and not the filling of the soul as a vessel, meerly by pouring into it from without, but the kindling of it from within; or helping it so to excite and awaken, compare, and compound its own notions, as whereby to arrive at the knowledge of that which it was before ignorant of.

(In this passage Cudworth for the moment leaves the question unanswered whether the ideas are innate in us, but from note 3, p. 33 follows that Cudworth leads us beyond both assumptions: namely, that of the dependence of the ideas on the sensible world, the soul as tabula rasa, and that of the innate ideas. He overcomes the static nature and the sterility of both hypotheses by putting the whole emphasis on the activity and the actualisation of reason which is realised in participation in the divine mind, which itself is pure act.)

[2] Plato. Phaedr. 249 *e*–250 *a*.

and actually realised[1]. Therefore the soul must in every act of thinking transcend its own individuality and finiteness and rise up to the universality of the ideas[2]. From the high place of the Universals, reason comprehends the particular by lifting it up to the Universal[3]. Knowledge does not, therefore, begin in the particular, it is only stimulated by it. Knowledge begins with the turning away of the soul from the particular objects towards the ideas; knowledge actually ends in the particular by the subsumption of the particular under the Universal, and its comprehension within the Universal[4].

Whether the sensible world can become intelligible, as Cudworth declares in his theory of atomism, will depend on whether this subsumption of the particular under the Universal can ever be wholly achieved or not. It is clear that atomism actually asserts distinct knowledge of the sensible world by reducing its qualities to quantitative proportions.

Later on we shall find these two currents in Cudworth's thought diverge even more widely, and we shall then see why Cudworth considers atomism merely a hypothesis, although the best, indeed the only one, he thinks, which can save the intelligibility of the outside world.

[1] Mor. 251. ... the first intellect IS essentially and archetypally all RATIONES and verities, and all particular created intellects are but derivative participations of it, that are printed by it with the same ectypal signatures upon them.

[2] ibid. 232. Wherefore the apodictical knowledge of this truth is no otherwise to be attained than by the mind's ascending above sense, and elevating it self from individuals to the comprehension of the universal notions and ideas of things within it self, making the object of its enquiry and contemplation, not this nor that material individual triangle without it self, but the indivisible and immutable notion of a triangle.
Cf. Plato. Pheado 107*b*.

[3] Mor. 133.

[4] ibid. 218. ... it is most certainly true, that they (intellection and knowledge) proceed from a quite different power of the soul, whereby it actively protrudes its own immediate objects from within it self, and comprehends individuals without it, not passively and consequentially, but as it were proleptically, and not with an ascending, but with a descending perception; whereby the mind first reflecting upon it self, and its own ideas, virtually contained in its own omniform cognoscitive power, and thence descending downwards, comprehends individual things under them. So that knowledge doth not begin in individuals, but ends in them.

A different aspect of the virtually omniform cognoscitive power of the soul is seen, when we consider the content, more than the act, of knowledge. We then discover that this cognoscitive power includes in itself anticipations of all possible Being, of all ideas, which make it possible for ideas to be awakened in the soul[1]. The soul is not a tabula rasa, in which ideas, originating from sensible things, could be engraved.

Because ideas are universal, indivisible and of absolute precision[2], they cannot be derived from empirical things. They are, on the contrary, active energies in the soul. The idea of a circle, for instance, is universal[3]. It is identical at all times and in every finite act of thinking. Empirically it cannot be represented accurately, and even if it could, nothing would be gained; for, quite apart from the fact that our senses are not acute enough to apprehend absolute precision, we could not recognise a perfect representation, unless we possessed the anticipation of the perfect circle.

A perfect empirical circle would, moreover, be a particular circle and therefore finite and accidental. There could be derived from it no universal geometrical theorems.

Cudworth emphasises the fact that there is a difference in kind, not only in degree, between an empirical and imagined circle, and a circle conceived by pure thought[4]. Without the idea of the circle we should

[1] ibid. 185 ff.
ibid. 93 ff. For to know and understand a thing, is nothing else but by some inward anticipation of the mind, that is native and domestick, and so familiar to it, to take acquaintance with it ...
Plato. Meno 80*d–e*.

[2] Mor. 201–2. ... when we look upon the rude, imperfect and irregular figures of some corporeal thing, the mind upon this occasion excites from within it self the ideas of a perfect triangle ... whose essences are so indivisible, that they are not capable of the least addition, detraction or variation without the destruction of them.
T. I. S. 827. ... all the abstract essences ... are indivisible.

[3] Mor. 229. And if there were any mathematically exact, our sense could be no criterion or rule to judge of it, nor discern when any thing were indivisibly such, nor judge of the absolute and mathematical equality ...

[4] ibid. 204. ... there being a vast difference betwixt the confused indistinction of sense and fancy, by reason of their bluntness and imperfection, and the express accuracy, preciseness and indivisibility of those intelligible ideas that we have of a straight line ... and other geometrical figures; and therefore that imperfect, con-

have no standard, by which to judge empirical circles. Every empirical circle would then necessarily appear to us as perfect[1].

Since without ideas we should have no standard of perfection, so neither could we have any aesthetic judgment[2]. For, in the creation of beauty, art copies the perfect and gives it a sensible form in order to lead the spirit, which beholds it, towards the perfect itself.

The idea accordingly is rule, paradigma and standard for the finite Being, and for the knowledge of it[3]. Without ideas we could never pass beyond the fortuitousness of the particular in imagination to the universality and necessity of thought[4]. Without ideas science would not be possible at all[5].

Sensible things have logical content, and are therefore intelligible, in so far as they are able to represent their corresponding ideas[6]. But this can never be perfectly achieved[7], and things are never fully intelligible. Fortunately this is not necessary, because the function of sensible things is only to stimulate the mind to thought. True know-

fused indistinction of sense, could never impress any such accurate ideas upon the mind, but only occasion the mind actively to exert them from within it self.

[1] ibid. 206. Which latter busy anticipation of it is the rule, pattern and exemplar, whereby he judges of those sensible ideas or phantasms. For otherwise, if there were no inward anticipations or mental ideas, the spectator would not judge at all, but only suffer; and every irregular and imperfect triangle being as perfectly that which it is, as the most perfect triangle ...

[2] ibid. 208 ff. For there could be no such thing as pulchritude and deformity in material objects, if there were no active power in the soul of framing ideas of regular, proportionate and symmetrical figures within it self, by which it might put a difference between outward objects, and make a judgment of them.

[3] T. I. S. 734 (by mistake 728. 734 and 735 are numbered as 728). ... a mind before the world, and senior to all things, no ectypal, but archetypal thing, which comprehended in it, as a kind of intellectual world, the paradigm or platform, according to which this sensible world was made.
(In this passage Cudworth comes nearest the neoplatonic static conception of νοῦς which principally he has overcome.)

[4] ibid. 735.

[5] ibid. 736.

[6] Mor. 163.
Plato. Tim. 31*a*.

[7] Mor. 89, 200. (The idea of the triangle cannot be represented empirically.)
Plato. Ep. 7, 343*a*.
Plato. Phaedo, 74*a*. (The things cannot by a long way reach up to the ideas.)

ledge is found by the mind when it acts independently of things and in perfect recollection within itself. We do not need absolute knowledge since, in the fortuitousness of the empirical world, we could apply it as little as the ideas can be adequately represented in particular things. Our problems are practical and finite, and they do not therefore require a final and absolute solution[1].

Ideas are thus not merely the logical content of things, but also their Being[2]. So far as things succeed in representing their corresponding ideas they ARE in their Becoming; yet because they never perfectly reach up to the idea, they never attain to an incorruptible and true Being, but can only ever and ever become towards Being.

The idea is Being, meaning and purpose of the things. It is their final cause, towards which they are becoming[3]. Being, intelligibility, and perfection go hand in hand also in the things[4], and thus it becomes clear that ideas have priority over things, and that reason has a 'natural imperium' over the sensible world[5].

[1] T. I. S. 639. ... yet may rational souls frame certain ideas and conceptions, of whatsoever is in the orb of being, proportionate to their own nature and sufficient for their purpose.
(This passage is the clearest proof that Cudworth was well aware that we can actualise also pure conceptions only in an inadequate manner.)
Cf. Plato, Theaet. 155*d*; Phaedo, 75*a*; Soph., 254*a*.

[2] Mor. 275. ... and whatsoever is clearly conceived or understood, is an entity.
ibid. 189. Which relative ideas being not comprehended by sense, and yet notwithstanding, the natures of all compounded corporeal things, whether artificial or natural, that is, whether made by the artifice of men or nature, consisting of them.

[3] T. I. S. 681 ff. (A house is not defined by the description of the material used for it, nor by the labour of the workmen, apart from the plan and the determination of its purpose.) .
ibid. 682. ... things of nature, in whose very essence final causality is as such included.

[4] ibid. 639. ... where there is more of light, there is more of visibility, so where there is more of entity, reality and perfection, there is more of conceptibility and cognoscibility.
ibid. 830. ... and since there is unquestionably, much more of entity in life and cogitation, than there is in meer extension or magnitude, which is the lowest of all being, and next to nothing.

[5] ibid. 844. ... cogitation is in order of nature, before local motion, and incorporeal before corporeal substance, the former having a natural imperium upon the latter.
ibid. 858. Mind and understanding hath a higher degree of entity or perfection in it, and is a greater reality in nature, than meer senseless matter or bulkie extension.

Finally, remembering what has been said about substances and the atomistic approach to clear and distinct knowledge of the sensible things, we can see, how much on the outskirts of Cudworth's philosophy are those things which he took from Descartes, and how very loosely they are connected with his deeper thought.

In spite of the fundamental difference between imagination and thinking, there is still a manifold interplay between them. Ideas can incorporate themselves into imaginative pictures[1], and thus keep the imaginative power occupied, while reason proceeds independently and recollectedly in pure thinking. This happens often, for instance, in mathematical thought, but also in language, in speaking or writing[2]. A great part, in any case the more ordinary part, of all communication between men, rests upon the possibility that spirit can incorporate itself in the word. It reveals, on the other hand, that corporeal vehicles are but symbols, which at best can but serve pure thinking. An adequate expression of the spirit in sensible things can never be expected, because spirit and matter are altogether incommensurable[3]. Cudworth suggests that, in this sense, we may even speak of books as having body and soul. In our communication with each other, we depend entirely on the capacity of spirit to read spirit, when incited by the humble garments of exterior signs[4]. A more adequate embodi-

[1] Mor. 212 ff. And so there is a phantasm and a conception at the same time concurring together, an active and a passive cogitation. The conception or intelligible idea being as it were embodied in the phantasm, which alone in it self is but an incomplete perceptive cogitation of the soul half awakened, and doth not comprehend the indivisible and immutable notion or essence of any thing.

[2] ibid. 185 ff.

[3] ibid. 216. ... nature doth as it were import various sentiments, ideas, phantasms and cogitations, not by stamping or impressing them passively upon the soul from without, but only by certain local motions from them, as it were dumb signs made in the brain.
cf. 1st Sermon, p. 6. Neither are we able to inclose in words and letters the life, soul and essence of any spirituall truths; and as it were to incorporate it in them.
(Therefore there can be no mechanical transference of truth from one person to another.)

[4] ibid. 40–1. The best written truth remains closed for us ... until we have a living spirit within us, that can decypher them: until the same spirit, by secret whispers in our heart, do comment upon them, which did at first indite them.

There is a CARO and a SPIRITUS, a flesh and a spirit, a bodie and a soul, in all the writings ... but there is a soul, and spirit of divine truths, that could never

ment of the spirit, than dead letters, are living deeds; but to this we shall come later.

Perfect communication of thought between men, therefore, can never be attained, for it would presuppose that spirit could freely shine into spirit, without the help of any mediating links, and this is unattainable for us, as long as we still need corporeal means of expression for our communication. The one exception of a more perfect communication is given us in pure thought, where at all times and in every finite mind, exactly the same ideas are remembered through participation in the perfect mind of God.

Aesthetics

Every representation of spirit in matter, every incarnation of the divine life, is beauty. It is of the very essence of beauty to bear the seal of reason and understanding in the harmonies of its proportions[1]; whether it be the seal of the divine mind of the creator, manifest in the harmonies of the universe, or of the finite mind of an artist in his work[2]. The most perfect beauty is found in the harmonies of logic, mathematics, and the ethical harmonies of holiness[3].

Contemplation of beauty is an essentially reasonable act[4]. Beauty is not a fact, neither does it give itself at man's will. The immediate

yet be congealed into inke, that could never be blotted upon paper which by a secret traduction and conveiance, passeth from one soul unto another; being able to dwell or lodge no where, but in a spirituall being, in a living thing; because it self is nothing but life and spirit. Neither can it, where indeed it is, expresse it self sufficiently in words and sounds, but it will best declare and speak it self in actions.

[1] Mor. 160 ff. For what is pulchritude in visible objects, or harmony in sounds, but the proportion, symmetry, and commensuration of figures, and sounds to one another, whereby infinity is measured and determined and multiplicity and variety vanquished and triumphed over by unity.

Plato, Tim. 69*b*.

[2] Mor. 170. For there is a nature in all artificial things, and again, an artifice in all compounded natural things.

[3] Mor. 181. (There are also ethical harmonies.)

... The man will also espy some symbolical resemblances of morality, of vertue and vice in the variously proportioned sounds and airs.

[4] ibid. 177–8. (The universe) ... and so as it were resounds and reechoes back the great creator's name, which from those visible characters impressed upon the ma-

vision of harmony is an awareness of the supreme order of the Whole[1].

When a man and a beast hear the same melody, the latter may hear its separate sounds, but the spirit of man alone, by means of the anticipation in its creative cognoscitive powers, hears the "together" of the sounds, "the one whole"[2].

Just as each act of thinking is, in a sense, a repetition in a finite mind of the pure act of the divine mind, so the contemplation of beauty is a repetition of the original creative act of the artist. It is also an essentially creative act, and depends on the creative strength of him who contemplates. The purer and the more differentiated are his anticipations[3], the greater is his comprehension of beauty. The object of aesthetic contemplation and pure thought is fundamentally the same: the one divine order and purpose of the universe, fixed for all time. The difference lies in the manner of apprehension. In contemplation we apprehend it in one intuition, while discursive reasoning, with much labour, seeks to work out, one by one, the relations which constitute the whole harmony.

Sensations, which are mere passions of the soul, are given us for the stimulation of aesthetic contemplation, just as sense perception stimulates to thought. They are not given for our deception. It is not, however, the senses but the spirit, which perceives beauty and rejoices, when, in the contemplation of beauty, it recognises its own and recreates it in itself. The spontaneous sensation of joy and delight which may accompany the contemplation of a work of art is not its final purpose, but only the first lighting up of the true vision of beauty and the first upspringing of the pure joy of the spirit[4].

terial universe, had pierced loudly into its ears, but in such an indiscernable manner, that sense listening never so attentively, could not perceive the least murmur or whisper of it.

[1] ibid. 178. But the unity of the whole harmony, into which all the several parts conspire, must needs proceed from the art and musical skill of some one mind, the exemplary and archetypal cause of that vocal harmony, which was but a passive print or stamp of it.

[2] ibid. 180.

[3] ibid. 183 ff.

[4] ibid. 180–1. ... when it finds or meets with in sensible objects any footsteps or resemblances thereof, any thing that hath cognation with intellectuality; as pro-

The most sublime object of aesthetic contemplation are the harmonies of the universe, for they reveal the spirit of its divine creator. They who consider the universe as derived from the arbitrary motion of dead matter, and they who, like Descartes, think of it as a perfect machine, which is set going by God at the start, look at the divine work of art as do animals, without reason and thought[1]. Yet Pan still plays the flute, but only the spirit hears the sound of his song[2].

Interaction in Knowledge

In the sphere of Being Cudworth believed he had found a solution for the problem of the interaction between the substances in the hypothesis of Plastic Nature, which, in intimate vital union, methodically leads and directs matter from inside towards its final cause. The same problem arises in the theory of knowledge. How can the essentially different faculties of sense perception and reason act upon each other? So far as Cudworth holds to the Cartesian dualism of substances, he also takes refuge in the hypothesis of the mediating function of the "esprits animaux". But he admits that this does not really convey much more than different expressions for nature's 'magic'[3].

portion, symmetry and order have, being the passive stamps and impresses of art and skill (which are intellectual things) upon matter, it must needs be highly gratified with the same.

cf. Plato, Nom. 667*d*.

[1] T. I. S. 684. Now as this argues the greatest insensibility of mind, or sottishness and stupidity, in pretended theists, not to take the least notice of the regular and artificial frame of things or of the signatures of the divine art and wisdom in them, nor to look upon the world and things of nature with any other eyes, than oxen and horses do.

[2] Mor. 184. ... but sense, which only passively perceives particular outward objects, doth here, like the brute, hear nothing but meer noise and sound and clatter, but no musick or harmony at all; having no active principle and anticipation within it self to comprehend it by, and correspond or vitally sympathize with it; whereas the mind of a rational and intellectual being will be ravished and enthusiastically transported in the contemplation of it; and, of its own accord, dance to the pipe of Pan, nature's intellectual musick and harmony.

cf. Plato, Phaedr. 250*a*.

[3] Mor. 82. ... secretly instructed by nature...

The dualism of substances leads altogether into a blind alley. But the striking fact is, that just there, where Cudworth is endeavouring to follow Descartes most closely, he found his own original approach. And now at last we come to the centre of his thought.

The decisive struggle between his dogmatic and critical tendencies shows itself in his vacillation concerning the inference of the reality of the outside world. Is this inference an inference of reason, or a "bastard" inference (*νόθος συλλογισμός*)[1]? As an atomist, Cudworth professes, almost dogmatically, that Extension with its modifications exists outside the mind, that it is absolutely intelligible, and that the certainty of its reality is arrived at by a conclusion of reason.

As a critical thinker, on the contrary, Cudworth admits that all Being is Being of the idea, and that we can know the sensible world only as far as it represents the ideas. In other words we can conceive only its logical content. There is now no need for any magical action, for spirit comprehends spirit: the spirit of the thinker divines the idea in the appearance of the sensible things.

We realise at once that the sensible world could never be wholly understood, even if its full logical content were known. We are therefore forced to an indirect inference of something existing outside reason as a deficient cause of knowledge (cf. *νόθος συλλογισμός* in Timaeus)[2]. This deficient cause cannot be grasped by reason, and therefore no predication whatever can be made of it. We must admit, that we know nothing at all of what it is. From this it follows, that we have access to the sensible world only as far as we comprehend and think it in ideas, because we have no other means to attain to knowledge,

[1] T.I.S. 48.

[2] cf. ibid. 638–9. For it is certain, that we cannot fully comprehend our selves, and that we have not such an adequate and comprehensive knowledge of the essence of any substantial thing, as that we can perfectly master and conquer it. It was a truth, though abused by the scepticks, that there is *ἀκατάληπτόν τι*, something incomprehensible, in the essence of the lowest substances. For even body it self, which the atheists think themselves so well acquainted with, because they can feel it with their fingers, and which is the only substance that they acknowledge either in themselves or in the universe, have such puzzling difficulties or entanglements in the speculation of it, that they can never be able to extricate themselves from.

than the ideas. That which evades the grasp of reason remains obscure and closed, not only to reason, but also to our senses.

This same ambiguity in Cudworth's philosophy appears in the theory of underlying substances.

As a dogmatic thinker Cudworth attributes to Extension an independent substantial Being outside reason, which remains untouched by the changes of its accidents, and is altogether beyond the process of becoming.

Besides Extension, he assumes the incorporeal substance Soul, which remains identical with itself, and indifferent to any process of becoming, and therefore indifferent also to the changes of its own consciousness.

In sharp contrast to all this we have his critical view. As a critical thinker Cudworth interprets body as essentially in the flux of Becoming. As far as it has Being, it has it from the idea. That means that it has Being in the degree, in which it becomes towards its idea, and realises its logical content in itself.

Likewise, Soul has no substantial Being in itself. It has Being only in so far as it activates itself in ever new participations in the divine mind. Since it awakens in itself ideas only, when incited by experience, it is itself, together with the increase of knowledge, becoming and IS not. Only in so far as it participates in the pure act of the divine mind has it existence in itself.

That which is becoming has no stability in itself whatever. From moment to moment its Being falls partly away into Not-Being. To finite beings, Becoming of existence and knowledge is essential and not accidental. Finite mind, in its progressing experience and knowledge never remains the same, because every newly attained knowledge modifies its entire content of thought, by putting its constituent parts into new relations[1].

We must now ask whether any certainty of Truth can be attained on such critical foundations. This leads us to the problem of the criterion of Truth.

[1] ibid. 720. But created beings, have but a derivative participation hereof, their understanding being obscure, and they erring in many things, and being ignorant of more.

Chapter III

Part I

THE CRITERION OF TRUTH

**What is Truth? – Cudworth gives two answers:
Truth IS. – Truth is Life.**

Truth IS. Truth is not created[1], and there is in Truth no process of Becoming, but it is absolute, necessary, eternal and immutable[2]. It IS all the essences of the things, the ideas and their necessary and unchangeable relations. Truth is the truths of all things. It is this as one single thought, one divine order and purpose, one perfect sphere of coherent meaning, one catholic Truth[3], the Absolute[4].

[1] ibi . 718. Truth is not factitious; it is a thing which cannot be arbitrarily made, but IS.

[2] Mor. 250. ... the immediate objects of intellection and science, are eternal, necessarily existent and incorruptible. ... there is an eternal wisdom and knowledge in the world, necessarily existing, which was never made, and can never cease to be or be destroyed; or, which is all one, that there is an infinite eternal mind necessarily existing, that actively comprehends himself, the possibility of all things, and the verities clinging to them.

[3] ibid. 258. ... truths, though they be in never so many several and distant minds apprehending them, yet they are not broken, multiplied, or diversified thereby; but they are one and the same individual truths in them all. So that it is but one truth and knowledge that is in all understandings in the world. Just as when a thousand eyes look upon the sun at once they all see the same individual object. Or as when a great crowd or throng of people hear one and the same orator speaking to them all, it is one and the same voice that is in the several ears of all those several auditors; so in like manner, when innumerable created understandings direct themselves to the contemplation of the same universal and immutable truths, they do all of them but as it were listen to one and the same original voice of the eternal wisdom that is never silent; and the several conceptions of those truths in their minds, are but like several echo's of the same VERBA MENTIS of the divine intellect resounding in them.

[4] Mor. 270–1. Whenever any theoretical proposition is rightly understood by any one particular mind whatsoever, and wheresoever it be, the truth of it is no

In every search after knowledge, even in every doubt, we presuppose this one divine order and purpose of coherent meaning. As ultimate foundation of every knowledge, it is always already believed, while it is yet sought as the end.

Truth is Life

Truth is the *νοῦς* (the divine mind) itself, which in one pure act, without deficiency, comprehends itself[1], the measure of its perfection, its creative power and communicative goodness[2]. The *νοῦς*

private thing, nor relative to that particular mind only, but it is a catholic and universal truth... throughout the whole world, nay, it would not fail to be a truth throughout infinite worlds if there were so many, to all such minds as should rightly understand it.

[1] T. I. S. 846. ... an absolutely perfect mind ... doth not (as Aristotle writeth of it) *ὁτὲ μὲν νοεῖν, ὁτὲ δὲ οὐ νοεῖν*, sometimes understand, and sometimes not understand; it being ignorant of nothing, nor syllogising about any thing; but comprehending all intelligibles, with their relations and verities at once within itself; and its essence and energy, being the same.

ibid. 737. ... such a mind as is essentially act and energy; and hath no defect in it. And this ... can be no other than the mind of an omnipotent, and infinitely perfect being, comprehending it self and the extent of its own power, or how far it self is communicable, that is, all the possibilities of things, that may be made by it, and their respective truths.

[2] Mor. 251. It is all one to affirm, that there are eternal RATIONES, essences of things, and verities necessarily existing, and to say that there is an infinite, omnipotent and eternal mind, necessarily existent, that always actually comprehendeth himself, the essences of all things, and their verities; or, rather, which IS the RATIONES, essences, and verities of all things; for the RATIONES and essences of things are not dead things, like so many statues, images or pictures hung up somewhere by themselves alone in the world: neither are truths meer sentences and propositions written down with ink upon a book, but they are living things, and nothing but modifications of mind or intellect; and therefore the first intellect IS essentially and archetypally all RATIONES and verities, and all particular created intellects are but derivative participations of it, that are printed by it with the same ectypal signatures upon them.

T. I. S. 857. So is the first mind or understanding, no other, than that of a perfect being, infinitely good, fecund and powerfull and vertually containing all things; comprehending it self and the extent of its own goodness, fecundity, vertue and

thinks the ideas of all possible Being, that means of everything, which is not contradictory in itself, thinks it actually from eternity. It is the spirit which neither slumbers nor sleeps, the sun which never goes down, the ultimate transcendental foundation of every finite act of thinking[1].

These two definitions of Truth do not contradict each other, for in perfect knowledge subject and object cannot fall apart, so that the known object is exterior to the knowing subject. Were it so, passivity, the very characteristic of imperfection, would still adhere to the pure act of thinking. The *νοῦς* as pure act brings forth its object ever new, not in time, but in eternity. It is itself act of knowledge and content of knowledge in one.

We are here at the central problem of philosophy and a passing glance at other systems may prove a help to understand and appreciate Cudworth's solution.

Hegel describes the spirit as subject and substance. As substance, he calls it "transparent stillness", where all the ideas are present in unperturbed transparence and limpid identity, when considered apart from the dialectical process.

But the spirit is always subject also, passing from thesis through antithesis to synthesis; it is bound in the infinite dialectical speculative process progressively to gain consciousness of its own content. Spirit is here also conceived as act and content in one inseparable unity, yet there is a fundamental difference between Cudworth and Hegel. Hegel

power; that is, all possibilities of things, their relations to one another and verities; a mind before sense, and sensible things. An omnipotent understanding being, which is it self its own intelligible, is the first original of all things.

[1] Mor. 255–6. ... since our understandings are but potentially all things, that is, have not an actual but potential omniformity only, there must of necessity be IN RERUM NATURA, another intellect that is actually all knowledge, and is the same to our understandings that active art it to passive matter and that the light is to our eyes, and which does not sometimes understand, and sometimes not understand, but is always eternal actual knowledge. A sun that never sets, and eye that never winks.

assumes that this process of self-consciousness of the spirit is accomplished in history, through the sense experience of finite instrumental minds. Cudworth, on the contrary, keeps to the fundamental and unbridgeable gulf between perfect and finite spirit, and he thereby escapes the tragic consequences of Hegel's conception, by which the spirit, in the infinite process of dialectical speculative progression, finally remains imprisoned in the finite substance, and itself sinks into finiteness, unable to regain its original transcendency. Thus, with Hegel, "transparent stillness" remains the transcendental presupposition of the act of thinking. It is also its final cause, but removed now into unattainable heights, to which the spirit can never return, when once immersed in the finite.

Kant. A reference to Kant presents us with a different aspect. Kant placed the transcendental foundation of the act of thinking, as well as the forms of thinking, the pure concepts (Die reinen Verstandesbegriffe), in the conception of the Unity of Apperception (Einheit der Apperzeption). It would be interesting to examine, whether, and if so, in what way, he assumed also for the content of knowledge a transcendental foundation. This leads further to the problem of how the Unity of Apperception is related to the DING AN SICH, whether, or to what extent, the conception of the DING AN SICH, as outside the Unity of Apperception, is already critically overcome by Kant himself. From this arises directly the central problem of Kant's philosophy, if and how the falling apart of the foundation of act and object of knowledge leads to the difference between Pure Reason and Practical Reason, and thereby to the two disparate certainties of the "I am certain" of Practical Reason, and the "It is certain" of Pure Reason; two certainties which are bound to restrict and, in part, exclude each other.

Truth is the *νοῦς* itself, as act and content of knowledge, but the criterion of Truth cannot therefore lie outside, but only within, the *νοῦς*[1], and the evidence of knowledge lies in the oneness of subject and object.

[1] T. I. S. 718. ... the measure and rule of truth concerning them, (the universal theorems of science) can be no foreign or extraneous thing, without the mind, but

Does this lead us any further? A thinking mind needs a criterion of Truth, when it is liable to doubt. But the perfect mind cannot doubt. It sends itself out into the multiplicity of its ideas, and, by the very same act, recalls itself into unity; unity of the one supreme order of coherent meaning (Sinnzusammenhang), as well as unity of the one pure act. The *νοῦς* itself immutably creates Truth in eternity; it therefore needs no criterion of Truth, for this could only mean a criterion of itself which would have no meaningful purport.

How does the problem of the criterion of Truth present itself in regard to finite knowledge?

From Cudworth's definition of Truth (Truth as the *νοῦς* itself), it follows, that all knowledge is knowledge of God himself[1]. God is himself, strictly speaking, the only subject of knowledge, since finite minds can activate themselves only by participation in him[2]. God is also the only object of knowledge, because he himself actually IS all things according to their logical content, which is their true Being.[3] And

must be native and domestick to it, or contained within the mind it self; and therefore can be nothing but its clear and distinct perception.

Mor. 280. ... the comprehension of that which absolutely IS; which is not terminated in the appearance only, as sense is, but in that which is, and whose evidence and certainty is no extrinsecal adventitious, and borrowed thing, but native and intrinsecal to it self.

cf. ibid. 271. Since the immediate objects of intellection exist in the mind it self, we must not go about to look for the criterion of truth without ourselves.

ibid. 272. ... the criterion of true knowledge is not to be looked for any where abroad without our own minds, neither in the heights above, nor in the depths beneath, but only in our knowledge and conceptions themselves.

T. I. S. 853. (An archetypal mind) that is senior to the world, and all sensible things, it not looking abroad, for its objects any where without, but containing them within it self.

[1] ibid. 733. (The genuine and original knowledge is the knowledge of God of himself.)

[2] T. I. S. 737. ... one only original mind ... all other minds whatsoever partaking of one original mind; and being as it were stamped with the impression and signature of one and the same seal. From whence it cometh to pass that all minds in several places and ages of the world, have ideas or notions of things exactly alike, and truths indivisibly the same.

[3] ibid. 767. ... the proper object of mind and understanding, is a perfect being, and all the extent of its power; which perfect being, comprehending it self and the

for us also the things are accessible only according to their logical content[1].

The problem of evidence is presented to us in the following way: can we, in a finite act of knowledge, as thinking subjects, ever attain the oneness with the object of our knowledge? If so, in what respect can we achieve it? This leads us to a renewed examination of finite knowledge.

The finite act of knowledge is not, as Descartes assumed, an actualisation of the universal substance Cogitation, but it consists therein, that a finite soul participates in a finite way in the divine *νοῦς* as act and content of knowledge[2]. This participation is manifest in that cognoscitive power, native to the soul, by which it is able to reproduce all the ideas, which the divine mind actually creates eternally. This it can do, not eternally, in a pure act of knowledge, but in time only, and successively[3].

Or, from the point of view that the *νοῦς* actually IS all things intelligibly, so IS the participating soul also all things, but virtually only. This means that the soul has a virtual omniformity; not however in the sense that it has at any time all the ideas at its command, as if it could draw freely on a hidden store of unconscious knowledge. The

extent of its own power, or the possibilities of all things, is the first original mind, of which all other minds partake.

Plato, Nom. 716.

[1] Mor. 130–1. Wherefore it must of necessity be granted, that besides passion from corporeal things, or the passive perception of sense, there is in the souls of men another more active principle or an innate cognoscitive power, whereby they are enabled to understand or judge of what is received from without by sense.

[2] Mor. 35–6. Now all the knowledge and wisdom that is in creatures, whether angels or men, is nothing else but participation of that one eternal, immutable and increated wisdom of God, or several signatures of that one archetypal seal, or like so many multiplied reflections of one and the same face, made in several glasses, whereof some are clearer, some obscurer, some standing nearer, some further off.

[3] ibid. 255. Now because every thing that is imperfect must needs depend upon something that is perfect in the same kind, our particular imperfect understandings, which do not always actually contain the RATIONES of things and their verities in them, which are many times ignorant, doubting, erring and slowly proceed by discourse and ratiocination from one thing to another, must needs be derivative participations of a perfect, infinite and eternal intellect, in which is the RATIONES of all things, and all universal verities are always actually comprehended.

ideas are not innate to the soul, neither are they simply given as facts[1], but the soul has a capacity to awaken them within itself by its participation in the divine mind. The soul must rise up to this participation by transcending its own finiteness in each act of knowledge[2].

The soul, moreover, depends on the experience of the senses in so far as the problems of the outside world urge it to thought[3]. Yet it can only find the solutions to these problems, when it turns away from the contemplation of the sensible things[4] and, recollected within itself, looks up to the ideas. From that high place, "with a commanding view", it returns to the particular phenomenon with which it started, subsuming it now under the Universal, thus making it transparent in its logical content[5]. But since experience is given

[1] ibid. 134–5. So all created intellects, being certain ectypal models, or derivative compendiums of the same; although they have not the actual ideas of all things, much less are the images or sculptures of all the several species of existent things fixed and engraven in a dead manner upon them; yet they have them all virtually and potentially comprehended in that one cognoscitive power of the soul which is a potential omniformity, whereby it is enabled as occasion serves and outward objects invite, gradually and successively to unfold and display it self in a vital manner, by framing intelligible ideas or conceptions within it self of whatsoever hath any entity or cogitability.

cf. ibid. 256. And from hence it comes to pass, that all understandings are not only constantly furnished with forms and ideas to conceive all things by, and thereby enabled to understand all the clear conceptions of one another, being printed all over at once with the seeds of universal knowledge, but also have exactly the same ideas of the same things.

(Such a passage as this may easily lead to the misinterpretation that Cudworth assumed innate ideas.)

[2] See Note 2. p. 35

[3] See Note 3, p. 29

[4] ibid. 137. ... scientifical knowledge is best acquired by the soul's abstraction from the outward objects of sense, and retiring into it self, that so it may the better attend to its own inward notions and ideas.

T. I. S. 732. ... the knowledge of this and the like truths is not derived from singulars, nor do we arrive to them in a way of ascent, from singulars to universals, but on the contrary having first found them in the universals, we afterwards descending apply them to singulars: so that our knowledge here is not after singular bodies, and secundarily or derivatively from them; but in order of nature, before them, or proleptical to them.

[5] Mor. 133–4. ... individual things existing without the soul, are but the secondary objects of knowledge and intellection, which the mind understands not by looking

us only successively, and since we are incapable of a survey of future experience, our knowledge of ideas also remains fundamentally incomplete[1].

In the pure act of knowledge Truth is activated as one immutable system of relations. Essentially different from this is the discursive act of knowledge, which can only proceed from one particular problem to the next, whereby unavoidably every new solution modifies the whole of the former knowledge. With the increase of knowledge ever new relations are formed, both between the objects of knowledge themselves, and between the objects and the thinking subject[2].

Only in the *νοῦς* would it be possible to give a final account of the pure concepts, but there such an account is not needed.

Finite knowledge is essentially a Becoming towards Being (*γένεσις εἰς οὐσίαν*).

Absolute Truth stands behind it as its presupposition and before it as its final cause. It itself moves in a sphere BETWEEN absolute knowledge and absolute ignorance, progressing from one finite problem to another, unable to lay hold on Truth as an absolute certainty[3].

It is, therefore, essentially impossible for a finite mind to attain evidence of the whole sphere of order and purpose (Sinnzusammenhang).

out from it self as sense doth, but by reflecting inwardly upon it self, and comprehending them under those intelligible ideas or reasonings of its own, which it protrudes from within it self; so that the mind or intellect may well be called the measure of all things.

[1] F.W. 40. And indeed the necessary understanding that is our clear conception and knowledge going so little way...

T.I.S. 882. ... so much more may the whole corporeal universe far transcend those narrow bounds, which our imagination would circumscribe it in.

ibid. 670. ... we mortals do all stand upon too low a ground, to take a commanding view and prospect upon the whole frame of things; and our shallow understandings are not able to fathom the depths of the divine wisdom, nor trace all the methods and designs of Providence.

[2] (This line of thought is not explicitly drawn out by Cudworth.)

[3] ibid. 874–5. There are indeed many things in the frame of nature, which we cannot reach to the reasons of, they being made by a knowledge far superior and transcendent to that of ours and our experience and ratiocination, but slowly discovering the intrigues and contrivances of Providence therein.

... wherefore we must not conclude that whatsoever we cannot find out the reasons of, or the use, that it serves to, is therefore ineptly made.

We may now ask further: can we obtain evidence at least in the solution of any particular problem? In other words, can we in any particular act of knowledge overcome the duality of subject and object?

We realise that this question implies a certain contradiction, since, strictly speaking, one cannot assume any degrees of evidence; knowledge is either evident or not. Perfect evidence implies insight into the whole divine order of ideas. In such knowledge as ours, incomplete and ever modified by new experience, evidence would necessarily be endangered by all new knowledge, even if it were possible in one particular solution. Evidence in finite knowledge can, strictly speaking, be obtainable only, when our sense experience of the outside world is finally completed.

However, we are bound to examine the question of evidence in particular problems, were it only because Descartes himself unhesitatingly asserted it. This examination will, moreover, show especially clearly, how the dogmatic and critical trends in Cudworth's philosophy continually meet and cross one another.

Can evidence be obtained in the solution of particular problems?

Cudworth distinguishes two kinds of knowledge[1].

a) Knowledge of necessary Being, i.e. of the first original objects of the spirit; the knowledge of pure ideas in their necessary relations (les vérités éternelles according to Leibniz[2]).

b) Knowledge of the sensible world, i.e. the secondary objects of the spirit (les vérités de faits according to Leibniz), presented to us by means of sense perception. We know them only in so far, as we succeed in subsuming them under the primary objects of reason. Science in the strict sense is possible only in the sphere of pure ideas[3]. The further an object is from the sensible world, the more clearly can it be known.

This leads us to the problem of the relation between the ideas and the sensible world.

[1] Mor. 102.

[2] ibid. 35. Science or knowledge is the comprehension of that which necessarily is.

[3] T.I.S. 731.

Plato. Tim. 27*d*–28*a*.

Plato. Res Publ. 508*d*–*e*.

The ideas are often called by Cudworth the Essences or Natures of things; he uses Idea, Essence and Nature interchangeably. In the relationship between Essences and sensible things he distinguishes two different aspects, in one of which the Essences are thought of as eternal, in the other as transitory[1].

a) The Essences as eternal. The ideas or Essences are the thoughts of God, functions of the *νοῦς*. They ARE all things according to their logical content[2]. They are indivisible like the numbers, and therefore identical with themselves, immutable and eternal. The least change would annihilate an idea as this particular idea and would cause it to be another idea[3].

In themselves the ideas are independent of finite thinking[4]. They would be the same, even if there had never been a finite mind, and they would remain the same, if at some time, there were to be no finite minds to think them. They are universal and in every finite act of thinking they remain identical at all times. But outside of the *νοῦς* they have no independent existence[5], for their very Being is their

[1] Mor. 284–5. ... there is an eternal knowledge and wisdom..., which comprehends within it self the steady and immutable RATIONES of all things and their verities, from which all particular intellects are derived and on which they do depend. But not that the constitutive essences of all individual created things were eternal and uncreated, as if God in creating of the world did nothing else, but ... only cloathed the eternal increated and antecedent essences of things with a new outside garment of existence, and not created the whole of them. And as if the constitutive essences of things could exist apart separately from the things themselves.

[2] T. I. S. 835. (The essences, as also the logical content of the sensible things, are conceived as functions of the eternal *νοῦς*.)

[3] Mor. 242. The essences of things are like to numbers; because if but the least thing be added to any number, or substracted from it, the number is destroyed.

[4] ibid. 248–9. Nay, though all the material world were quite swept away, and also all particular created minds annihilated together with it; yet there is no doubt but the intelligible natures or essences of all geometrical figures, and the necessary verities belonging to them, would notwithstanding remain safe and sound. Wherefore these things had a being before the material world and all participating intellects were created.

[5] ibid. 250. ... objective notions ... which are things that cannot exist alone, but together with an actual knowledge in which they are comprehended, they are the modifications of some mind or intellect.

BEING KNOWN. This applies to every Being; Being is Being of idea[1].

The sensible things are constantly "Becoming towards Being", but in the unceasing flux of the Becoming they cannot preserve their Being fully, they ARE only in so far as they are intelligible.

From this the assumption of a substance Extension as essentially different from Cogitation appears again under a new aspect. In accordance with this same definition of Being, Cudworth concludes the existence of Extension and its modifications from its intelligibility. But this is a deceptive accordance, in truth the assumption of a duality of substances is incompatible with it. If Being is intelligibility, there can be outside and apart from the idea absolutely no independent substantial Being, not even in the realm of the finite.

The ideas are also the true Natures of things. Not only as their Being, but also as their meaning and purpose, are they the final causes of things, towards which these are moving in the flux of their Becoming, and by which alone they can be comprehended[2]. The final cause is therefore the only true cause of things. All mechanical causes are merely CONDICIONES SINE QUIBUS NON.

If the ideas are the final causes of the Being, meaning and purpose of things, they naturally cannot be completely severed from them, nor remain in an inaccessible transcendency, but they must have reality in the things. Indeed, things essentially ARE their intelligible final causes, and are essentially constituted by the coherent system of their interrelations[3]. That does not, however, mean that the ideas are immanent in the things[4]. Yet the ideas in the divine *νοῦς* are, as the transcendental causes of the things, also their meaning, purpose and Being.

[1] T. I. S. 718. ... whatsoever is clearly perceived to be, IS.

ibid. 736. ... though there be no absolute necessity that there should be matter or body, yet is there absolute necessity that there should be Truth.

[2] Mor. 168 f.

Plato, Phileb. 16*d.*

[3] Mor. 157.

[4] ibid. 270. Truth is the most unbending and uncompliable, the most necessary, firm, immutable, and adamantine thing in the world.

b) The essences as transitory. If the essences are conceived as directly and materially constituting the things, they must have been created together with them, and must perish with them, they cannot therefore be eternal[1].

Although this distinction is not very clear, it nevertheless shows, that Cudworth endeavours at all costs to uphold the unbridgeable transcendency of the idea; though, on the other hand, he safeguards himself against the assumption that the Essences are independent self-existent entities, static SUBSTANTIAE SEPARATAE which, as he says, would represent some kind of intelligible spectres.

According to Cudworth, the Neo-platonists were guilty of this absurd travesty of true Platonism, which inevitably issued in idolatory, and the worship of ideas as divinities.

A further reason against the hypostatic subsistence of ideas is, that the creative act of God cannot consist in merely clothing with existence Essences, already in Being. God creates the whole essence of things, and their actuality implies their existence. He creates them new and entirely out of himself[2], according to the idea in his mind, which, as a paradigma, precedes and determines every act of creation, and which is for all created things at the same time the ultimate end, from which it has Being, meaning and purpose. Thus the idea is the transcendental presupposition and 'vocation' for both Being and knowledge of everything.

Knowledge, distinct from sense perception and imagination, is essentially knowledge of the final cause, of "the whole"[3]. The parts of a house, for instance, are only understood in their final cause and in the relation-system of their specific purpose. It is the design, after which a house is built, which constitutes it as a house, and distinguishes it from a heap of stones. The arrangement of the particular

[1] T. I. S. 835.

[2] ibid. 750. (That souls are created out of nothing means that they have their whole being and existence from God alone.)

[3] Mor. 164. Wherefore the eye of sense of this self-mover, though it be fixed never so much upon the material outside..., yet it never comprehends the formal nature of it within it self, as it is a whole made up of several parts, united not so much by corporeal contact and continuity, as by their relative conspiration to one certain end.

parts of the material is only its CONDICIO SINE QUA NON. Or a watch IS primarily the relation-system of the different wheels as 'one whole', and not the specific kind and position of the particular wheels as such[1]. This system of coherent meaning and purpose, "the whole", is essentially reasonable and can be seen and understood by reason alone.

"The whole", the meaning and purpose, the proportion of things, is their beauty[2]. Thus every act of knowledge is fundamentally an aesthetic vision, and every aesthetic vision is an act, of which only the spirit is capable. A work of art, whether it be the universe, the work of the artist God[3], or a human work of art, presents to us the world, or one part of it, as one perfect order of meaning and purpose, as a harmony. In contemplating it, we repeat the creative act of the artist by virtue of the anticipations in our own mind. The mind, awakened by contemplation[4], self-actively realises in itself the relation-system of the specific ideas as one whole.

The vision of beauty is the direct lighting up in the spirit of the one divine and perfect order of meaningful purpose, whether it be incited by contemplation of the primary objects of the spirit, the logical, mathematical, or ethical harmonies, or whether it be mediated by sensible representation[5]. The multiplicity of the particular parts is

[1] Mor. 162.

[2] ibid. 160. For what is pulchritude in visible objects, or harmony in sounds, but the proportion symmetry and commensuration of figures and sounds to one another, whereby infinity is measured and determined, and multiplicity and variety vanquished and triumphed over by unity?

[3] Mor. 175.

[4] ibid. 175. Wherefore the ideas of art and skill, author and artificer were not passively imprinted on the intellect from the material self-mover, but only occasionally invited from the mind it self, as the figures of the engraven letters did not passively impress the sound of the artificer's name upon him, but only occasion him to exert it from his own activity.

ibid. 176–7. For being ravished with the contemplation of this admirable mechanism and artificial contrivance of the material universe, forthwith it naturally conceives it to be nothing else but the passive stamp, print and signature of some living art and wisdom; as the pattern, archetype and seal of it, and so excites from within it self an idea of that divine art and wisdom.

[5] ibid. 181.

then overcome and taken up into unity in "the one whole"[1]. Moreover, in the vision of beauty, we overcome the falling apart of subject and object, and thus attain to a certain oneness with the object, for then we can no longer say, whether the contemplated object is in us, or we in it[2].

Evidence in particular knowledge could thus be obtained, if only this oneness did not break as soon as, in regaining self-consciousness, we reflect upon the object seen. With this breaking apart evidence also is lost.

Cudworth nevertheless asserts that in contemplation we have a more direct way of knowledge which leads us further than discursive reasoning[3]. Yet he admits that the further we advance in it, the more formidable grows the discrepancy between the things we apprehend and our possible expression of them.

Is there an analogous possibility of evidence in thinking?

Cudworth, as we have seen, divides knowledge into two different spheres, knowledge of pure ideas and knowledge applied to sensible things.

Knowledge of pure ideas is actualised in pure logic, pure mathematics, and in the foundations of ethics. The mind, by participation in the divine *νοῦς*, awakens in itself the immutable ideas and their necessary relations. Knowledge of pure ideas is not dialectical, and therefore evidence in this sphere could be obtained. Cudworth actually assumes such a sphere of certainty of knowledge, though he

[1] See Note 2 on previous page.

[2] ibid. 98. ... the intellect doth read inward characters written within it self and intellectually comprehends its object within it self, and is the same with it.

(Cudworth here assumes oneness with the object, and therefore evidence also for the particular dianoetic act of thinking. This passage shows that every act of knowledge is in his view an aesthetic vision, and that the sensible world can be apprehended in and through the ideas.)

[3] T. I. S. 781. Now the reason why we cannot frame a conception of such a timeless eternity, is only because our selves are essentially involved in time, and accordingly are our conceptions chained, fettered, and confined to that narrow and dark dungeon, that ourselves are imprisoned in; notwithstanding which, our freer faculties, assuring us of the existence of a being, which far transcendeth our selves, to wit one that is infinitely perfect; we have by means hereof, *μαντεΐαν τινά*, a certain vaticination, of such a standing timeless eternity, as its duration.

regards it as extremely limited[1], both upwards and downwards[2]. Here again, only one small middle section is accessible to us, that of the philosophical and mathematical axioms[3].

Logical axioms give us no content; whenever we try to apply them to any given fact, their evidence is immediately lost.

The upward limitation of evidence is the knowledge of God. When God is the object of knowledge this oneness could be obtained, at least as far as the object is concerned, because God is perfectly intelligible and there is no sub-logical part in him. The obstacle lies in ourselves, in the limited capacity of our finite reasoning. With God as the object of knowledge, we can never be one, because the infinite is incommensurable with us, who are finite. We have, it is true, a certain anticipation of the existence and of the attributes of God, but an adequate knowledge of him we can never attain[4]. Demonstrations

[1] Fr. W. 41. The weakness of human understanding is such that there are very few things which men do so certainly know as that no manner of doubt may be raised in their minds against them.

[2] Mor. 271–2. The necessary truth of no geometrical theorem can ever be examined, proved, or determined by sensible things mechanically. And though the eternal, divine intellect be the archetypal rule of truth, we cannot consult that neither, to see whether our conceptions be commensurate with it.

[3] T. I. S. 720–1. And it seems to be no derogation from Almighty God to suppose that created minds by a participation of the divine mind, should be able to know certainly ... common notions, which are the principles from whence all their knowledge is derived. And indeed were rational creatures never able to be certain of any such thing as this at all; what would their life be but a meer dream or shadow? And themselves but a ridiculous and pompous piece of phantastick vanity?

... knowledge is the only thing in the world, which creatures have, that is in its own nature firm; they having here something of certainty, but no where else.

Wherefore we having now, that which Archimedes required, some firm ground and footing to stand upon, such a certainty of truth in our common notions, as that they cannot possibly be false; without which nothing at all could be proved by reason.

Cf. Plato, Tim. 49*e* and 53*d*.

[4] T. I. S. 643. As we have no phantasm of any infinite, so neither is infinity fully comprehensible by our humane understandings, that are but finite.

ibid. 640. ... every thing is apprehended by some internal congruity in that which apprehends ... Wherefore it cannot possibly otherwise be, but that the finiteness, scantness and imperfection of our narrow understandings, must make them asymmetral or incommensurate, to that which is absolutely and infinitely perfect.

of the existence of God lead no further than to the knowledge, that God cannot be denied by logical argument, for he himself is the presupposition, object, and end of every act of thinking.

The downward limitation concerns the sensible world. There the realm of necessary knowledge is limited by the nature of the sensible things. The obstacle comes from the side of the object, which is inadequate to the idea and contains an unintelligible sub-logical part[1]. Therefore the application of the ideas to the explanation of sensible things necessarily becomes dialectical, tending always to dissolve "evidence" into doubt, and every new piece of knowledge demands a new examination of that already known.

That evidence cannot be obtained in this sphere is thus due primarily to the character of the object of knowledge, but also to the fact that the experience of the thinking subject is never completed.

The middle sphere, between the "upward and downward" limitation, is the sphere of the pure ideas, of logic, mathematics and ethics. Even in this middle sphere, difficulties arise. We have seen, that we awaken the ideas within us, when we are incited by the problems of sense experience[2]; and that further, we can never have the whole realm of ideas present in our minds at once, and therefore can never be sure, whether, by new experience, new pure ideas and new axioms

[1] T. I. S. 780, 736. (Cudworth only hinted at the fact that the individual can never be reached by thought nor by the atomistic reduction of the sensible world into intelligibility, nor by the ideas. Even if the atomistic hypothesis of the reduction of sensible things into clearly and distinctly intelligible quantities is assumed and if all the sensible qualities are ascribed to the soul alone as its mere sensations; this is still no more than a formal statement and reveals nothing of the content of the sensations.)

[2] Mor. 93. For otherwise reason alone without sense could not acquaint us with individual existent things without us, or assure us of the existence of any thing besides God, who is the only necessarily existent being.

(Cudworth here makes the interesting restriction that we are dependent on experience for the knowledge of everything whatever except for the existence of God since this is necessary in itself. We regret that Cudworth never examined the question whether the logical and mathematical axioms also can be known by pure thought independently of experience. Such an examination might have brought out still more clearly his theory that the intellect can only activate itself when it is incited by experience.)

will not be remembered in us. With this uncertainty the evidence again becomes conjectural[1].

Cudworth nevertheless believes he has found, at least in this sphere, an Archimedial point, where knowledge, independent of the finite subject, is in itself evident and where, as he puts it, all knowledge comes to rest.

From that Archimedial point we cannot, however, derive a system of sciences through inference according to analogy of evidence, as Descartes attempted. In the first place, ideas are not at our command, since we depend on the incitement of experience, which in its turn, is never completely under our control. We are therefore altogether incapable of making a complete and conclusive deduction. Secondly, an evident system of sciences cannot be built up, before the experience of the sensible world is completed, since the knowledge of facts is dialectical. The third argument against a complete system of sciences is, that evident knowledge of facts can never be attained at all.

All the logical axioms are founded upon the incompossibility of contradictions. Cudworth lays such a great emphasis on the evidence of the principle of contradiction, because in it he sees the presupposition of all knowledge[2].

Absolute evidence of axioms, which unfortunately never gives us any evident content, cannot convey more than a bare certainty of the Absolute. Or, to put it in negative terms, by logical argument it cannot be denied, that there IS Absolute Truth. Yet evidence of axioms never leads to the absolute certainty of any content of knowledge.

From this it follows that we can attain to no evidence in any particular knowledge. We have no choice, but to commit ourselves, at

[1] T. I. S. 639. Truth is bigger than our minds, and we are not the same with it, but have a lower participation only of the intellectual nature , and are rather apprehenders than comprehenders thereof. This is indeed one badge of our creaturely state, that we have not a perfectly comprehensive knowledge, or such as is adequate and commensurate to the essences of things.

[2] Mor. 32. ... the compossibility of contradiction destroys all knowledge and the definite natures or notions of things.

the risk of error, to a certain clearness of apprehension [1], without any absolute standard being possible. The divine νοῦς is too high, and the sensible world too low, for us to measure and verify by them our knowledge.

This compromise is similar to that made by Cudworth, when he assumed that we normally have a certain practical assurance and judgment in sense perception, even if we have once been deceived by it, although strictly speaking, after one deception, certainty is for ever lost.

From what we have seen, it is not surprising to learn that Cudworth, in spite of his sound critical examination of the problem of evidence, speaks often in the context of the theory of substances, of clear and distinct knowledge, adopting Cartesian terminology, although in the core of his thought he has proved evident knowledge to be impossible.

Part II

CRITIQUE OF CARTESIAN PRINCIPLES OF KNOWLEDGE

Cudworth's critique of the metaphysical foundation of the Cartesian philosophy is a most admirable part of his work. Not so much, perhaps, as regards the actual critique of Descartes, for he probably

[1] ibid. 278–9. ... the ultimate resolution of theoretical truth, and the only criterion of it, is in the clearness of the apprehensions themselves, and not in any supposed blind, and unaccountable make of faculties.

ibid. 282. And it cannot be denied but that men are oftentimes deceived, and think they clearly comprehend what they do not: but it does not follow from hence, because men sometimes think that they clearly comprehend what they do not, that therefore they can never be certain that they do clearly comprehend any thing; which is just as if we should argue, that because in our dreams we think we have clear sensations, we cannot therefore be ever sure, when we are awake, that we see things that really are.

Mor. 138. ... the mind is naturally conscious of its own active fecundity, and also that it hath a criterion within it self, which will enable it to know when it hath found that which it sought.

did not do full justice to him, and we can hardly assume, that he read Descartes fully or even sufficiently. The great value of this critique is the clarity with which it brings out the power and beauty of his own thought, and shows the significance of his placing the criterion of Truth in the objectivity of the absolute and perfect mind.

I will give first a short sketch of the Cartesian principles of knowledge, as far as it is necessary for the understanding of their refutation.

Descartes' thought is based on principles of Geometry. The universe presented itself to his mind as an intelligible system of timeless proportions. A perfect mind could comprehend it in one single glance, but we, with our finite cognoscitive powers, have to proceed by discursive reasoning, by the imperfect method of deducing one thing from another. We cannot advance with certainty, unless we have found the one Archimedial point, on which the whole system of proportions can, with certainty, depend. This Archimedial point must be directly evident in itself, absolute and unassailable by any present or future doubt. In the search for it Descartes proceeds by the rule of the methodical doubt: "to acknowledge as true only that which cannot possibly be subjected to doubt; that which by our means of knowledge is comprehended as beyond doubt."

The whole realm of knowledge must therefore be subjected to this test; whatever part should prove subject to possible doubt must be excluded as unfit for the foundations of knowledge.

The methodical doubt begins with:

1. Sense perception. The argument runs thus: We do not perceive the same object always in the same way. For instance, things appear different from afar than from near; a square tower may appear round in the distance, or objects, when held in water, are perceived to be of a different shape. A staff appears broken, when dipped in water. This proves that sense perception can deceive us and must therefore be excluded from being a reliable principle of knowledge.

2. Yet there is still the possibility that we perceive the smallest parts of the object correctly, but combine the various perceptions in a wrong way. In this case the mistake is in thought, not in perception; and the task would be to analyse sense perception into its smallest element and to find the correct combination. But this process is re-

stricted by the fact that below a certain minimum our capacity of perception fails. And even if we could achieve this analysis completely, we should still not attain any certainty, because

3. we perceive objects with equal clearness, when awake and when dreaming. In both conditions we have the impression of experiencing reality and therefore lack the criterion of whether we are awake or dreaming. This doubt cannot be solved directly. We must therefore seek a kind of knowledge, which is unaffected by it, by being equally true, when we are awake or dreaming. Thus we arrive at

4. the logical and mathematical truths. They remain valid, whether we doubt the reality of the sensible world or not, and are exactly the same, whether we dream or wake. 2+2 is always 4. It seems, that here we have reached a point, where the doubt ends. But alas, a more formidable one immediately arises.

5. We have no criterion for the integrity (veracity) of our cognoscitive powers. It is possible to doubt them, for a "mauvais génie" might have created us with faculties, which necessarily conceive clearly and distinctly falsehood as truth. In this case we are inevitably always deceived, even in that, which seems to us perfectly intelligible.

For this doubt there is no solution. But one thing is established: it is absolutely certain that, when I doubt, I am thinking, and when I think I must exist.

I. Cogito, sum (I think, I exist). This certainty of my existence at the moment of doubt seems beyond the realm of doubt altogether, since it sprang directly from it. But no continuity of my existence as yet is guaranteed. Only the certainty that, whenever I shall doubt again, I shall exist. However, only in the moment of doubt do I need this certainty. Yet as long as the integrity (veracity) of our cognoscitive powers remains questionable, we cannot trust even this direct certainty. To meet this, Descartes argues that, if the existence of a perfect Being could be proved, this formidable doubt would be overcome. Perfection implies perfect goodness and omnipotence. It would contradict the goodness of a perfect creator to deliver his creatures to inevitable deception, and his omnipotence would be denied by the assumption, that a demon could spoil his creation against his own will.

II. The Demonstration of the existence of God. While I doubt, I acknowledge myself as an imperfect Being, but thereby I testify, that I bear within me the idea of perfection, by which I measure my own imperfection. This idea of perfection cannot have originated from myself. It must have been put into me by an actually existent perfect Being, without whom I should have no standard of perfection. Man, thrown upon himself, would necessarily believe himself perfect; but if the idea of a perfect Being is in him, God must actually exist, and his existence guarantees once for all the integrity (veracity) of our cognoscitive powers.

These two certainties, the certainty of my own existence during the doubt, and the certainty of the existence of a perfect Being, are both absolute, because they both arose from a situation of most acute doubt. They are necessarily true, because their contrary cannot be thought. But in no way are they deducible one from the other and are to be understood as a proportion.

III. The ultimate purpose of the methodical doubt is, as we have seen, to lay a sure foundation for scientific research. This, of course, involves the existence of the outside world, which must be proved. Descartes attains to this last absolute certainty by the argument that the outside world must exist, or I could not doubt it.

These three certainties are self-evident and absolute. As a proportion they form the foundation of every existence, knowledge and Truth. The Archimedial point is found, from which, by analogy of evidence, every further knowledge can be deduced by the rule that knowledge is absolutely certain, if it is thought with the same necessity as these fundamental certainties. The criterion of Truth thus lies in the clearness and distinctness of each particular act of knowledge in a particular individual, which is a mode of the substance Cogitation.

We will now follow Cudworth's refutation of each stage of the methodical doubt, the three metaphysical certainties and the theory of evidence.

The refutation of the methodical doubt

Cudworth objects to the methodical doubt as such. He considers it as an altogether unsound method of thinking. For Descartes evidently claims absolute validity for it, even while purposing to doubt of everything[1]. Apart from this it cannot be acknowledged as a genuine doubt at all, precisely because it is methodical and presupposes, that we can attain to absolute knowledge, which in fact is the direct purpose of the doubt itself. In the study of the principles of ethics, we shall find this objection confirmed from quite a different point of view. Descartes safeguarded himself against any fatal consequences of the doubt in practical life by establishing preliminary moral maxims. This is, in Cudworth's view, a plain contradiction in terms. Genuine doubt arises from the perplexities of practical life, and we have no protection against it. It is practical experience, which presents the problems to the intellect to work upon, and which rouses the doubts which reason must solve. Without them we should have no need to think at all.

This same objection can be made against *Hegel*, when he asserts that the antithesis is built up by the spirit itself, in order that the spirit, by finding the synthesis, may advance in its own selfconsciousness.

Genuine doubt can never be absolute doubt, because experience, as well as thought, proceeds only as we pass from one problem to another. We can never doubt everything at once, for this would imply that all things, capable of being objects of experience, were presented to us at once. Absolute doubt presupposes the possibility of a completed experience, and of absolute knowledge. But in any pretence of absolute doubt, we necessarily deceive and contradict ourselves, because, even if we reject every certainty, we are still forced to claim a certain dogmatic validity for all our negative assertions[2].

Cudworth's second objection is still more serious. We have already seen, that he himself follows the first and second step of the methodical doubt, the doubt of the objective validity of sense perception and of the lack of a criterion, whether we are awake or dreaming. But he vigorously objects to the doubt that our cognoscitive powers might be

[1] Ibid. 273.

Ibid. 275. For unless they did acknowledge that we do clearly understand some things, they do not undertake so much as hypothetically to demonstrate any thing.

[2] Mor. 67.

made for error[1]; the doubt on which, in Descartes' system, the foundation mainly depends. Cudworth denies that Descartes has solved it, and adds that, if this doubt were ever admitted at all it could never be overcome by any finite being. And if, per impossibile, it were overcome, nothing would be gained. He then proves that the problem is falsely put, and that this doubt springs from altogether false metaphysical foundations of knowledge.

The refutation of the doubt concerning the cognoscitive powers

Descartes believes the solution of this doubt to lie in the demonstration of the existence of God. Cudworth does not acknowledge this, but objects, that Descartes goes round in a circle, and this he considers a rather serious lapse[2]. For, with as yet unassured cognoscitive powers, Descartes proves the existence of God, and from this demonstration, which itself is exposed to every deception, he presumes to derive the guarantee of the veracity of those same powers, by which he conducted the demonstration. It is, however, probable that Descartes considered the certainty of the existence of a perfect Being as a direct certainty, which precedes logical mediation with an a priori evidence, and which has a constructive import for the cognoscitive powers. But this is a questionable assumption, and Cudworth expressly denies, that a certainty can be attained, which is not mediated by thought, but is, as it were, placed beyond reason.

[1] T.I.S. 717. So that if we will pretend to any certainty at all, concerning the existence of a God, we must of necessity explode this new sceptical hypothesis, of the possibility of our understandings being so made, as to deceive us in all our clearest perceptions, by means whereof, we can be certain of the truth of nothing, and to use our utmost endeavour to remove the same.

[2] Ibid. 717. ... at once to prove the truth of God's existence from faculties of reason and understanding and again to prove the truth of those faculties, from the existence of a God essentially good; this I say is plainly to move round in a circle; and to prove nothing at all: a gross oversight, which the forementioned philosopher (Descartes) seems plainly guilty of.

Wherefore according to this hypothesis, we are of necessity condemned to eternal scepticism.

Mor. 274. Wherefore upon this supposition, all created knowledge as such is a meer phantastical thing.

Whenever this doubt is admitted, we have no possible way out of it, and at once there appears the danger of ultimate scepticism [1].

We can never get beyond our cognoscitive powers [2]. Our own thought can never be the direct object of our knowledge, because it is necessarily preceded by a thinking subject [3]. Everything, of which we can become conscious at all, must pass through thought, and God alone knows its integrity [4]. To prove such an integrity would presuppose that we could pass beyond our thought, but this can never be, and therefore absolute certainty in any knowledge can never be attained either.

From another aspect the situation appears still more precarious and scepticism even more imminent. Our faculties are accidental and created, therefore finite and imperfect. If Truth were founded upon them, then Truth itself would necessarily be hypothetical and lack universality and necessity. It would be relative to faculties, which are accidentally created, thereby becoming itself accidental. Even if the integrity of these faculties could be proved, they could still never be the agents of absolute Truth [5]. If Truth is made relative, the doors to all scepticism at once spring open.

Descartes believed that, with cognoscitive powers guaranteed once for all, he could pave the way to an autonomous, unlimited scientific research, for which God would still be, somehow, terminus a quo, and, in a certain sense, also the foundation of knowledge, but no more ter-

[1] T.I.S. 717.

Mor. 274.

[2] T.I.S. 720. ... we can go no further than our faculties; and whether these be true or no no man can ever be certain.

[3] Ibid. 733. (The act of thinking itself can never be the direct object of knowledge.)

But what account can we then possibly give, of knowledge and understanding, their nature and original? Since there must be νοητόν, that which is intelligible, in order of nature before νόησις, or intellection.

[4] Mor. 273. Our faculties ... of which none can have any certain assurance, but only he that made them...

[5] Ibid. 280–1. For if knowledge have no inward criterion of its own, but the certainty of all truth and knowledge depend upon an arbitrary peculiar make of faculties, which is not a thing knowable in it self, neither can there be any assurance of it given but what is extrinsecal by testimony and revelation (...), there will be no such thing as knowledge, but all will be meer credulity and belief.

Cf. Plato Phaedo 79*d*.

minus ad quem. Cudworth naturally resists such a claim of such autonomous thought, and considers it a dangerous illusion. Being finite, our cognoscitive powers can never be guaranteed, because ideas are not innate in us, as Descartes assumed, and are not at our command. We must look up to them and in every act of thinking, transcend anew our own finiteness and rise up to the participation in the universal NOUS. This demands ever new effort, or, as Cudworth says, diligent watchfulness of the spirit[1]. In every act of thinking, the strength of the cognoscitive powers is put to the test afresh[2]. They attain clearness of knowledge in the degree, in which they overcome their own finiteness and participate in the NOUS (νοῦς). This is primarily a moral act, and therefore the power of thinking is proved to depend on the general habitus of the thinker[3]. Descartes escaped this demand, but at the cost of losing the ground for the knowledge of Truth.

Cudworth shows, that any doubt concerning cognoscitive powers, is based on a contradictory conception of knowledge and Truth, that it is not a legitimate doubt, because it presupposes that falsehood could be clearly and distinctly known[4], and that God could have created cognoscitive powers for error.

That which is clearly and distinctly intelligible has objective reality in itself. It is, therefore, necessarily true. Being is being-known[5].

[1] Fr. W. 73. As we do more or less intend ourselves in consideration and deliberation, and as we do more or less fortify our resolutions to resist the lower appetites and passions, so will the appearances of good and our practical judgments be different to us accordingly.

Plato, Phaedo 79*d*.

Plato, Phaedr. 249*c*.

[2] Mor. 279. So that the certainty of clear apprehensions is not to be derived from the contingent truth of faculties, but the goodness of faculties is only to be tried by the clearness and distinction of apprehensions.

(Critique of reason must ACCOMPANY every particular act of knowledge.)

[3] T. I. S. 662.

[4] Mor. 275–6. And if it be absolutely impossible even to omnipotence, that contradictories should be true together, then omnipotence it self cannot make any such faculties as shall clearly understand that which is false to be true, since the essence of falsehood consists in nothing else but non-intelligibility.

[5] Ibid. 275. Nay, the true knowledge or science which exists no where but in the mind it self, has no other entity at all besides intelligibility; and therefore whatsoever is clearly intelligible, is absolutely true.

But that, which is in itself contradictory, cannot be comprehended[1], not even omnipotence could make it intelligible. An almighty will can only create Being by calling into existence the Possibilia. When God creates new Beings, he realises ad extra the content of his own mind, the Possibilia of things. His mind IS Truth itself, and, at the same time, criterion of Truth. It IS absolute certainty. To create false cognoscitive powers would mean creating them with the faculty of comprehending Not-being as Being. This is contradictory, and therefore nonentical. It would further imply that the will of God could realise itself independently of his wisdom, and with this all Truth and knowledge of Truth would be still more grievously endangered.

That cognoscitive powers could differ in their fundamental constitution, not merely in the clarity of their apprehension[2] would necessarily involve the compossibility of contradictions, whereby all Truth would be made arbitrary and subject to chance. Absolute certainty of knowledge is attained only by a mind which, bearing in itself the criterion of Truth, is in no respect relative to our accidental and created cognoscitive powers[3]. Yet, in a still more formidable way, Descartes subjects Truth to chance, when he assumes that it is created by an arbitrary act of God; and against this Cudworth opens his broadest batteries, as we shall see in the next chapter.

[1] T.I.S. 718. ... no power, how great soever, can make anything indifferently to be true.

[2] Ibid. 718. In false opinions, the perception of the understanding power it self, is not false, but only obscure.

Mor. 277. And though intellectual faculties may be made obscure more or less, yet it is not possible that they should ever be made false, so as clearly to apprehend whatsoever is true to be false, and what is false to be true.

[3] Ibid. 277. ... we ought not ... to suspect that truth and knowledge were such whiffling things, as that they meerly depended upon an arbitrary make of faculties.

Mor. 279. For be these faculties what they will, clear intellectual conceptions must of necessity be truths, because they are real entities.

Ibid. 280. Nay, to make the certainty of all truth and knowledge not to be determined by the clearness of apprehensions themselves but a supposed unaccountable truth and rectitude of faculties, and so by the uncertainty thereof, quite to baffle all our clearest intellections, is quite to pervert the nature of knowledge.

The refutation of the Cogito, sum

The question, whether self-consciousness can, as the Archimedial point, be the foundation of the certainty of existence and knowledge depends on the solution to the following problems:

a) Are we justified in considering a finite thinking subject as a substance of Cogitation?

b) May the substance of Cogitation be legitimately restricted to self-consciousness?

c) Is it possible to prove that the soul, as substance of Cogitation, thinks in a continuous, uninterrupted and uninterruptible progression of thought, always conscious of the whole range of consciousness? If, on the contrary, it should be proved that reason, both as regards the act of thinking and the content, is accidental, then certainty of Truth can never be founded on it.

The AUTEXOUSION. Cudworth arrives at self-consciousness by a different way, than that of the methodical doubt, to which he referred as unsound. He says that we are conscious of ourselves in two different manners[1], *a)* by the "inner sense"; and *b)* by means of the logical axiom that "nothing can come out of nothing"[2].

a) We have in our "inner sense", before any logical mediation, an immediate certainty of "I, myself", as an intending, perceiving, thinking and acting agent.

b) According to the axiom, that nothing can come out of nothing, we conclude, from the fact that we have life, understanding and will, that there is an indivisible and incorporeal "I, myself", which is

[1] Fr. W. 71. We are certain by inward sense that we can reflect upon ourselves and consider ourselves, which is a reduplication of life in a higher degree.

Ibid. 58–9. But whatever be the case of brute animals as to this particular, whose insides we cannot enter into; yet we being in the inside of ourselves do know certainly by inward sense that there is in us someone hegemonical, which comprehending all the other powers, energies and capacities of our soul, in which they are recollected and as it were summed up, having a power of intending and exerting itself more or less, determines not only actions but also the whole passive capability of our nature one way or other, either for the better or the worse.

[2] T. I. S. 637. We are certain of the existence of our own souls, partly from an inward consciousness of our cogitations, and partly from that principle of reason, that nothing can not act.

capable of acting upon itself, and which, as self-active source of its own motion, is also source of thought and will, and of all the lower faculties of the soul[1]. It is never given us as a factum and, as such, it cannot be the object of direct knowledge, since it precedes every act of thinking, and can only indirectly be deduced from its effects.

Thus far the "I, myself" is a transcendental cause, which nevertheless is finite, because its field of action is limited, and because it can act self-actively only by virtue of its participation in the divine *νοῦς*. This autexousion is the focus of energy in the soul, which explains the faculty of the soul to act upon itself, symbolised by Cudworth as a reduplication of the soul, in which it acts, as if it "took itself into its own hands"[2].

Every action of the soul ad extra is also an action upon itself, because every change in its consciousness implies a change in the relation of the soul towards itself. Therefore action upon itself must be attributed to the soul as an essential faculty[3]. It may be asked, whether Descartes, in founding the assurance of existence upon self-consciousness, had in mind an explicit self-consciousness, or whether he (as Cudworth understands it), meant only, that in each act of thinking the rational soul acts upon itself.

In the controversy with Thomas Hobbes, Cudworth wrongly rejects the view that the soul in every action acts at the same time upon the whole universe, since, with every change in itself, the relations to everything in the universe are of necessity altered.

[1] Ibid. 846. The rational soul is it self an active and bubbling fountain of thoughts; that perpetual and restless desire, which is as natural and essential to us, as our very life, continually raising up and protruding, new and new ones, in us; which are as it were offered to us. Besides which we have also a self recollective power, and a power of determining and fixing our mind and intention upon some certain objects, and of ranging our thoughts accordingly.

[2] Fr. W. 39. (The hegemonic of the soul) which is the soul itself comprehensive, and having the conduct and management of itself in its own hand.

Ibid. 36. (The hegemonic) and indeed that which is properly we ourselves, (we rather having those other things of necessary nature than being them), is the soul comprehending itself, all its concerns and interests, its abilities and capacities, and holding itself as it were in its own hand, as it were redoubled upon itself, having a power of intending and exerting itself more or less, in consideration and deliberation.

[3] Ibid. 71. ... for all cogitative beings as such are self-conscious.

The explicit action of the soul upon itself is achieved in the following ways:
By the intending of objects, by which the soul anticipates and determines sense perception.
By awakening of imaginative pictures within itself.
By reflecting upon its own content of consciousness.
By self-determination in indifferent and different choice.
By the proposing and pursuing of purposes.
All these actions upon itself are purely cogitative and energetic. They are explicable only as springing from an indivisible and incorporeal centre of energy[1].

Yet the action of the soul upon itself must not be restricted to the sphere of self-consciousness only. Cudworth considers it a fundamental mistake, that Descartes left no room for a rational cause, which acts unconsciously[2], a cause which methodically directs all vegetative life towards a final cause, thus forming a connection between Cogitation as self-consciousness and Matter as Extension, and which, in vital union with matter, directs its motion from within by cogitative energy[3].

Cudworth deplores the devastating consequences of Descartes' mistake in deeming local motion to be explicable by mechanical principles without a final cause, thus handing it over to mere chance and unintelligibility[4]. Nor did he stop there, but proceeded also to interpret

[1] T. I. S. 831. (The incorporeal substance) ... such a kind of entity, as hath an internal energy; acteth from it self, and within it self and upon it self.

Cf. T. I. S. 852. ... since there is such a thing as errour, or false judgment, all cogitations of the mind cannot be meer passions; but there must be something of self-activity in the soul it self, by means whereof, it can give its assent, to things not clearly perceived, and so err.

[2] Ibid. 175. (The Cartesian philosophy) ... in rejecting all Plastick Nature, it derives the whole system of the corporeal universe from the necessary motion of matter, only divided into particles insensibly small, and turned round in a vortex, without the guidance or direction of any understanding nature.

[3] Cf. ibid. 672.

[4] Ibid. 682. And no more certainly can the things of nature (in whose very essence final causality is as much included) be either rightly understood, or the causes of them assigned, merely from matter and mechanism, or the necessary and unguided motion thereof; without design or intention for ends and good. Neither indeed can we banish all final, that is all mental, causality from philosophy or the consideration

the vegetative life of animals as mechanical, and so prepared the way for atheism[1]. From Cudworth's hypothesis, that Being is Being of the idea, it is impossible to sever life and Cogitation from each other. If life is explicable from arbitrary motion without a final cause, there is nothing to prevent us from deriving all Cogitation from the same origin, dead matter[2].

In a different connection Cudworth points out that, if sensitive soul is suffered to fall apart from rational soul and denied immortality, no sufficient reason can be found for the assertion of immortality of the rational soul either[3].

As the principle of motion, the soul is the principle of all life, as well as of thought. Matter has no autokinetic power, but all local motion has a cogitative cause. If each body is constituted by local motion of atoms, it is as such in strict dependance on a cogitative cause, and therefore Extension cannot be considered co-ordinate with Cogitation[4]. The cogitative cause, as the only possible rational cause, ne-

of nature, without banishing at the same time reason and understanding from our selves; and looking upon the things of nature, with no other eyes, than brutes do.

Ibid. 382. ... to assert mind to be the maker of the world, is really all one, as to assert final causality for things in nature, as also that they were made after the best manner.

Ibid. 383. ... the philosophick humour of some in this present age, who pretending to assert a God, do notwithstanding discard all mental and final causality, from having any thing to do with the fabrick of the world; and resolve all into material necessity, and mechanism; into vortices, globulae and striate particles.

[1] Ibid. 687. We cannot otherwise conclude concerning these our mechanick theists, who will thus needs derive all corporeal things from a dead and stupid nature, or from the necessary motions of senseless matter, without the direction of any mind, or intention for ends and good; but that they are indeed cousin-germans to atheists; or possessed in a degree, with a kind of atheistick enthusiasm, or fanaticism; they being so far forth inspired with a spirit of infidelity, which is the spirit of atheism.

[2] Ibid. 763. For as much as this prepares a direct way to atheism; because if life and sense, cogitation, and consciousness, may be generated out of dead and senseless matter, then might this well be supposed the first original of all things.

[3] T. I. S. 44. See p. 18, note 1.

Ibid. 846. (As far as Descartes considered the animals as automatons, he conceived them as deprived of cogitation and true life. Therewith Cudworth does justice to Descartes; yet the Cartesian hypothesis cannot be valid, because every local motion has a cogitative cause.)

[4] Ibid. 849. ... were there once no life nor mind at all, these could never have been produced out of matter altogether lifeless and mindless.

cessarily acts towards one system of related meaning and purpose. This proves that no mechanism can exist disconnected from its final cause, but neither can knowledge of the true cause be gained on merely mechanistic suppositions[1].

If once the soul exists unconscious of the content of its consciousness, self-consciousness cannot belong to it of necessity[2]. Such is the case, for instance, in deep sleep or conditions of unconsciousness. The soul has, moreover, energies which it activates without direct consciousness, for instance, reflexes, instincts, habits, and organic functions, such as breathing, the beating of the heart and the assimilation of food. Of a vast field of the soul's action we know nothing at all; for example, its effects upon the body, the mystery of vital union, and the activities of the imaginative powers are almost entirely beyond our grasp.

A continuous and uninterrupted self-consciousness of the soul of its own content cannot be proved, either waking or sleeping. We have, therefore, no sufficient evidence for regarding the soul as the substance of Cogitation, if this is restricted to self-consciousness.

Cudworth puts forth even more weighty arguments: Whether logically mediated or not, self-consciousness is never more than accidental; and as a finite act of thinking it can present only a momentary accidental certainty of our existence as cogitative beings. It can never ensure the continuity of our existence; although we need this certainty only, when we are in doubt, the continuity of existence in the future is still vitally important, because, without it, we cannot propose purposes nor pursue them. Since every realisation of Cogitation is directed towards a final cause, the setting of purposes is essential to its every act, and therefore the continuity of existence is its inevitable presupposition.

If Descartes were right, the substance of Cogitation would have to realise itself in an uninterrupted connection of thought, both awake

[1] Ibid. 682. See p. 73, note 4.

[2] Ibid. 160. Now if the souls of men and animals be at any time without consciousness and self-perception, then it must needs be granted, that clear and express consciousness is not essential to life.

Ibid. 161. ... since no man can affirm that he is perpetually conscious to himself, of that energy of his soul, which does produce it when he is awake, much less when asleep.

and asleep[1]. To such a mind an accidental sensible experience could not become conscious; and without the outside world, the soul would have no possibility of acting at all; it would be deprived of the faculty of directing the attention towards certain objects, of intending future knowledge and of recalling to memory past thoughts.

A similar prospect opens up in the system of Hegel, where the spirit is assumed to gain self-consciousness through the dialectical process, which is achieved in a necessary linear progression of knowledge from thesis, through antithesis, to synthesis. Caught in such a linear progress, the spirit cannot go back to former knowledge, neither can previous knowledge be modified by any newly acquired results.

A further argument against continuous necessary progression of knowledge is that we have no store of ideas within us, which we can actualise independently of the stimulation of sense perception. We can only become conscious of ideas, when our mind, incited by experience, self-actively awakens them by virtue of its participation in the divine *νοῦς*. In us, therefore, the ideas are always accidental, not as regards their content, but as regards their being remembered at a certain time and in a specific connection of ideas of which they are part in our momentary consciousness[2]. Yet in so far as this momentary connection of ideas modifies also their content, they are accidental

[1] Fr. W. 27. Were our souls in a constant gaze or study, always spinning out a necessary thread of uninterrupted concatenate thoughts, then could we never have any presence of mind, no attention to passing occasional occurrences, always thinking of something else ... and ... so be totally inept for action.

Ibid. 61. If speculative and deliberative thought be always necessary in us, both as to exercise and to specification, then must it be either, because they are all necessarily produced and determined by objects of sense from without, or else because the understanding always necessarily works of itself upon this or that object, and passes from one object to another by a necessary series or train and concatenation of thoughts.

Ibid. 61 f. (In the latter case) ... then could we never have any presence of mind, no ready attention to emergent occurences or occasions, but our minds would be always roving and rambling out, we having no power over them to call them back from their stragglings, or fix them and determine them on any certain objects.

[2] T. I. S. 502. ... *νοήματα*, conceptions in the mind of God ... though not such slight and evanid ones, as those conceptions and modifications of our human souls are.

in us even as regards their content[1]. Therefore the soul, in its self-consciousness, can reflect only upon an accidental range of thought, whereby the very act of self-reflection is accidental too, since it is incited by experience.

A finite mind can never free itself entirely from fortuitousness; it cannot therefore be in itself substantial.

Since finite thought in general, and self-consciousness in particular, have proved to be accidental in act and content, it is impossible to found on them the absolute certainty of our existence. Universality and necessity, which alone can constitute certainty, could be found only in a pure act of thinking, which, as to the content, is absolutely independent of the outside world, and as to the act, independent of psychological dispositions of the thinker. But such perfect detachment from empirical conditions we can never attain[2].

An accidental act of self-consciousness can, therefore, not be the foundation of absolute Truth any more than accidental cognoscitive powers could be its guarantee.

The "Cogito, sum", does not lead us any further than to the accidental momentary certainty, that during an accidental act of thinking the existence of an actually thinking subject cannot be denied. Neither his existence, nor his character as substance of Cogitation, can possibly be guaranteed by it. Absolute certainty of our existence, of the existence of God and of the outside world, as fundamental presuppositions for knowledge, can never be attained by finite reason[3]. Absolute certainty can only be found in the objectivity of absolute Truth itself, which is necessary and self-existent and perfectly unaffected by any accidental finite properties.

[1] (Cudworth only mentions this consequence.)

Cf. ibid. 639. ... an idea of a perfect being, agreeable and proportionate to our measure and scantling.

See note 1, p. 38 and note 2, p. 43.

(This follows indirectly also from the refutation of Descartes' foundation of knowledge on guaranteed cognoscitive powers.)

[2] T. I. S. 725. ... whatsoever is received, is received according to the capacity of the recipient.

Ibid. Preface. (There is no absolute persuasive power against anti-moral interests, not even in geometrical definitions).

[3] Mor. 274.

Therewith the Archimedial point in Descartes' metaphysical foundation of knowledge becomes doubtful; since both the demonstrations of the existence of God, and the inference according to reason of the outside world, spring from accidental thinking, they cannot provide a metaphysical foundation of Truth. Finite reason itself is in strict dependence on its transcendental presupposition[1], which IS absolute Truth and the pure act of thinking. The nature of finite reason directly proves, that it can realise itself only upon the supposition of absolute Truth. It cannot therefore be its foundation[2].

The absolute certainty of the existence of God, gained by demonstration, cannot serve as a foundation for the knowledge of Truth. The actual form of the demonstration is, nevertheless, carefully examined, and discussed at length by Cudworth.

The same must be said of the certainty of the outside world. We have already, in connection with atomism, pointed out the peculiar significance given to the inference according to reason of the outside world in Cudworth's philosophy.

Evidence. Descartes' solution of the problem of evidence once more discloses the fundamental disagreement between him and Cudworth in the conception of knowledge. Descartes believed that in the first three metaphysical truths we have immediate evidence, which somehow precedes logical mediation. The content of Truth of any further knowledge can be measured by this same standard, in the sense that everything which is precisely as evident as the three metaphysical truths, is absolutely true. By virtue of our cognoscitive powers, once for all guaranteed by the demonstration of the existence of God, a way is opened to us to attain to absolute Truth; starting from the three self-evident absolute certainties, we can, by analogy of evidence, deduce a complete system of sciences.

Descartes conceives evidence as the psychological phenomenon of direct clearness of apprehension. It is founded in the subject itself;

[1] Ibid. 68.

[2] T. I. S. 737. Truths are not multiplied, by the diversity of minds that apprehend them; because they are all but ectypal participations of one and the same original and archetypal mind and truth.

... it is but one and the same eternal light, that is reflected in them all; ... or the same voyce of that everlasting word, that is never silent, reechoed by them.

but absolute certainty can, nevertheless, be inferred from it, because the soul is a mode of the universal substance of Cogitation, and because the cognoscitive powers are in themselves guaranteed to work correctly, independent of the moral habitus of the thinker.

Evidence of particular knowledge, even of knowledge of facts, if it is conceived as mere psychological fact, can indeed be attained. Therefore it seemed possible to Descartes to deduce a system of sciences with the force of evidence.

Descartes' theory of evidence is probably one of the weakest parts of his philosophy, yet the confrontation with Cudworth shows how directly it follows upon the start of his theory of knowledge in general. It reveals the tendency of his whole thought, which is to place the foundation of knowledge in the finite mind, in accidental cognoscitive powers, and in certainties attained by accidental acts of thinking.

We have followed Cudworth, step by step, in his refutation of Descartes' principles of knowledge, and with these falls also his conception of evidence.

After having shown, on what uncertain grounds Descartes' philosophy stands, Cudworth launches his heaviest charge against him, by proving that, on Descartes' supposition, there could be no absolute Truth at all, and that he, far from finding them, had destroyed the foundations of all knowledge of Truth.

Chapter IV

THE KNOWLEDGE OF GOD

Part I

THE KNOWLEDGE OF THE EXISTENCE OF GOD; THE DEMONSTRATIONS OF THE EXISTENCE OF GOD

Cudworth believes that true knowledge of God is not denied us completely, because "it would contradict the glory of God to have created rational beings who could not attain to the assurance of his existence"[1]. "God is a mountain", we cannot embrace it, yet we can touch it[2]. Cudworth sees three ways leading to the knowledge of God: discursive reason, contemplation and holiness. These three ways, in various manners, presuppose and cross each other. One cannot be taken without regard to the others. The knowledge of God demands the use of all our faculties, directly or indirectly, each one in its own way. It takes hold of our whole personality.

Discursive reason comprehends God by the demonstration of the existence of God. It comprehends him from behind, as it were, by his works[3] in space and time. We come nearer to the Truth when, with our freer faculties, we plunge into the "ocean" of God and lose ourselves in the contemplation of him[4]. Yet even so we behold only the smallest

[1] Ibid. 721. See p. 59, note 3.

... besides it is no way congruous to think, that God Almighty should make rational creatures so as to be in an utter impossibility, of ever attaining to any certainty of his own existence; or of having more than a hypothetical assurance thereof.

[2] Ibid. 639.

[3] Ibid. 474.

[4] Ibid. 640. ... which in the silent language of nature seem to speak thus much to us, that there is some object in the world, so much bigger and vaster than our mind and thoughts, that it is the very same to them, that the ocean is to narrow vessels, so that, when they have taken into themselves as much as they can thereof

part of him, because our vessels are very small. Moreover, as soon as we reflect upon that which we have "seen", trying to put it into words and communicate it to others, we have to use our narrower faculties of reason again.

The highest way to the knowledge of God is holiness[1]. There we gain an immediate certainty of him, because we participate in him in a more direct way. The perfect, which embraces all, can be realised in our finite life in one way only, when we do not exclude evil, but overcome it in love and transform it into good. Then we participate in God's own life so perfectly, that we no more know, whether we act in God or he in us[2]. From this participation in God springs an immediate certainty of his existence.

All these ways of knowledge are inadequate, because "Truth is bigger, than we are, and we are not identical with it"[3]. We are incommensurable with the perfect in thought, as well as in will.

by contemplation, and filled up all their capacity, there is still an immensity of it left without, which cannot enter in for want of room to receive it, and therefore must be apprehended after some other strange and more mysterious manner, viz., by their being as it were plunged into it, and swallowed up or lost in it.

Cf. Plato Phaedo, 85*d*.

[1] T. I. S. 203. ... there is a certain life or vital and moral disposition of soul, which is more inwardly and thoroughly satisfactory, not only than sensual pleasure, but also than all knowledge and speculation whatsoever.

[2] First Sermon, 5–6. All this is but the groping of the poore dark spirit of man after truth, to find it out with his own endeavours, and feel it with his own cold and benummed hands. Words and syllables which are but dead things, cannot possibly convey the living notions of heavenly truths to us. The secret mysteries of the divine life, of a new nature, of Christ formed in our hearts; they cannot be written or spoken, language and expressions cannot reach them; neither can they ever be truly understood, except the soul itself be kindled from within, and awakened into the life of them.

Ibid. 18–19. Knowledge indeed is a thing farre more excellent then riches, outward pleasures, worldly dignities, or any thing else in the world besides holinesse, and the conformity of our wills to the will of God: but yet our happinesse consisteth not in it, but in a certain divine temper and constitution of soul, which is farre above it.

[3] T. I. S. 639. See p. 61, note 1.

Ibid. 639. Nevertheless because our weak and imperfect minds are lost in the vast immensities and redundancy of the Deity, and overcome with its transcendent light and dazeling brightness, therefore hath it to us an appearance of darkness and incomprehensibility.

The demonstrations of the existence of God by discursive reasoning marks the least adequate way to the knowledge of God. They cannot, therefore, be given any central position in the theory of knowledge as is the case in the Cartesian metaphysics.

Cudworth's attitude towards the demonstrations of God is somewhat ambiguous. He attributes to them, on the one hand, a very great significance, when he, for instance, says that Descartes, by rejecting the cosmological demonstration of God, broke down the strongest defence against atheism. He is much concerned that religion, as well as knowledge, should be founded upon the perfect clarity of the divine mind, and not on the twilight of our subjective feelings of fear or hope[1]. Cudworth therefore searches for ever new arguments for the existence of God. On the other hand, Cudworth restricts the significance of the demonstrations of God to such a narrow range that it is difficult to see, why he devoted such a large part of his work to them. He says, for instance, that these demonstrations have no conclusive power in themselves at all. They are plausible only to a trained, alert and diligent mind; as soon as the attention is diverted, they unawares slide out of our hands. With good reason do some distrust such fine cobwebs of thought, for how could they possibly carry the sublime truth of the existence of God[2].

[1] Ibid. 664 sq. ... it might better be said, that the opinion of a God sprung from men's hope of good, than from their fear of evil; but really, it springs neither from hope nor fear... nor is it the imposture of any passion, but that whose belief is supported and sustained, by the strongest and clearest reason.

[2] Ibid. 725. ... men being generally prone to distrust the firmness and solidity, of such thin and subtle cobwebs or their ability to support the weight of so great a truth ... therefore shall we lay no stress upon this neither.

Ibid. 834. His (God's) existence being indeed demonstrable, with mathematical evidence, to such as are capable; and not blinded with prejudice, nor enchanted by the witchcraft of vice, and wickedness.

2nd Sermon, 196. The reasons of philosophy, that prove the soul's immortality, though firm and demonstrative in themselves, yet they are so thin and subtile to vulgar apprehensions, that they glide away through them, and leave no such palpable impressions on them as can be able sufficiently to bear up against that heavy weight of gross infidelity that continually sinks down the minds of men to a distrust of such high things as be above the reach of sense. Neither are these considerations any longer of force than men actually attend to the strength and coherence of the demonstration; and when that actual attention (which is operose and difficult) is

If the demonstrations of God were to be the ground for a demonstrative science of God, it would mean that God had to be a priori proved from antecedent causes[1]. God then would be denied self-existence, perfection and his very divinity. "Demonstration of God", in the strict sense of the word, is a contradiction in terms. The results of arguments for the existence of God are always indirect, they only show, that God cannot be denied by means of logic[2]. But absolute knowledge, mechanically conclusive as such, can never be attained[3].

Cudworth's chief opponents, Thomas Hobbes and the champions of Calvinism, even Descartes to a certain extent, all start by opposing reason to faith, but Cudworth shows that this separation is the principal source of many fatal errors, because it unavoidably ends in the assertion of a double Truth; or rather in the relativation of all Truth, either in the form of materialism, or of religious fanaticism, never far apart from one another. We have seen that Cudworth places the foundation of all knowledge and of all life in the one absolute Truth. In the face of this an irrational faith in God must necessarily appear contradictory and a mere nonentity. Revelation, which we could in no way grasp by thought, could not reach us at all, and would mean nothing to us. It could never become "for us". But even in itself it would be a nonentity, because all Being is founded on the divine *νοῦς*. Materialism and religious fanaticism, which latter Cudworth fights in the form of predestination to evil, both lack sound foundation, they are arbitrary and contradictory.

The great importance of the demonstrations of God is, that discursive reason proves the reasonableness of the faith in the existence of God. The existence of God is inferred by reason from our own life

taken off, then truth itself like a spectre or apparition suddenly vanishes away, and men question with themselves afterwards whether there were any such thing or no. Such thin and evanid things are philosophical speculations about the high mysteries of faith and religion.

[1] T. I. S. 715. ... the existence of God could not be demonstrated a priori, himself being the first cause of all things.

[2] Mor. 253.

T. I. S. 871.

[3] T. I. S. Pref. (It is impossible to put forward a convincing demonstration of God in the face of anti-ethical interests.)

and thought; as the first cause αἰτία, from which all Being has sprung, to which it is drawn back by the urge of an inborn impetus.

The conception of God. A quite disproportionate part of Cudworth's work is devoted to demonstrating that the idea of God is in man "by nature"; that it is present in rational Beings as such as an anticipation of a higher order, which in a certain sense embraces and implies all the other anticipations[1].

That does not however mean that faith in God can be derived from human reason, for this would lead no further than to the derivation of faith from feeling and imagination, because in finite reason too subjectivity and arbitrariness can never be fully overcome. "In the nature of rational Being" is equivalent, in Cudworth's argument, to "founded on the perfect νοῦς"[2], since only by participation in the νοῦς do rational Beings grow to what they are.

If the idea of God is present in rational Beings as such, it must indwell all rational Beings on earth. Actually to prove this, Cudworth makes an extensive research through all the world religions, accessible to him, and he arrives at the conclusion that in all religions the idea of one supreme, perfect, self-existent and necessary Being is to be found. These attributes, which are the unfolding of the one idea of perfection, can belong only to one single supreme Being[3], because two perfections would make each other relative. Therefore all religions, which believe in a perfect Being, are, of necessity, monotheistic[4]. True polytheism contradicts itself[5]. This also Cudworth found confirmed in the history of religion. The multitudes of gods, in whatever manner they may be conceived, are in all religions in a hierarchical order, ascending to the one supreme Being, on whom

[1] Ibid. Pref. The generality of mankind have constantly had a certain anticipation in their minds concerning the actual existence of a God according to the true idea of him.

Ibid. 665. ... there is also, besides a rational belief thereof, a natural prolepsis or anticipation in the minds of men.

[2] Ibid. 691. ... religion being founded, upon the instincts of nature, and upon solid reason.

[3] Ibid. 208. (The oneness of God follows from his attributes.)

[4] Ibid. 209 ff. (About genuine monotheism in all religions.)

[5] Ibid. 210. (True polytheism is necessarily contradictory.)

they all depend, and whom they serve each in his own particular field of action.

An appearance of polytheism often results, when the attributes and manifestations of the one supreme God are separately venerated, or when his creative power is worshipped directly in its manifestations on earth[1]. But even then, strictly speaking, the object of veneration always is the supreme Being.

True polytheism is professed only by a dualistic religion, where an evil god is believed to fight against a good God. This is a contradiction in terms and therefore furthest divorced from Truth; yet at the same time especially near to it, because such a view must have resulted from a sincere desire to keep far from God all responsibility for evil, and to believe unreservedly in his unchanging, perfect goodness alone[2].

Cudworth goes even further and says that, not only all religions, but even the atheists themselves, have the true genuine theistic idea of one perfect Being with its necessary attributes, or they could not deny it[3]. They profess, for instance, the idea of his perfect goodness, when they deny God, because of the wickedness of the world and his creatures. They acknowledge the idea of perfect justice, when they deny him on account of the bad dispensation of the world, and the apparent injustice reigning in it, which all too often lets the wicked go unpunished. Again, they have the idea of the perfect beatitude of God, when they consider it as being necessarily disturbed by everlasting anxieties about his creatures[4].

The belief in a universal anticipation of the existence of God in the human mind as such, leads us to a further aspect of Cudworth's philosophy. His theory of the cognoscitive powers and the virtual omniformity of the rational soul shows that he does not believe in a gradual development of the human mind through history. All knowledge was, either virtually or actually, present from the beginning. Cudworth is rather inclined to believe in a higher degree of knowledge and a deeper

[1] Ibid. 230. (Various possible forms of polytheism.)

[2] Ibid. 213 ff. (Religious dualism originates in the endeavour to save God from the responsibility of evil.)

[3] Ibid. 206.

[4] Ibid. 369.

participation in the mind of God in the most ancient times, which in the course of history passed through various phases of distortion and subsequent restoration; so, for instance, the doctrine of the divine Trinity, which Cudworth found hidden in nearly all religions; and the atomistic physiology, first discovered by the Pythagoreans, then distorted by Democritus, and others, and finally again restored by Descartes.

The fact that the idea of the existence of a perfect Being is so universal serves Cudworth as an argument for its reasonableness or, for what was to him the same, its being natural[1].

Demonstrations of the existence of God

The existence of God can be derived

1. From the idea of God. The fact that we have the idea of God in our mind testifies implicitly to his actual existence. The soul as a finite cause (αἰτία) is incapable of creating new ideas out of nothing, or forming an idea of a nonentity, that is, of something not actually and eternally realised by the eternal *νοῦς*. That which is not possible has no Being. If God, whose idea we have in us, were not at a certain time, he could never be, because a perfect self-existent Being cannot begin to be[2]. His existence would then be impossible. God, therefore, either IS, or he is impossible. If God did not exist, the soul would have formed the idea of an impossible Being, and this is inconceivable.

[1] T.I.S. 631.

[2] Ibid. 695. Many of those partial conceptions contained in the entire idea of God, are no where else to be found in the whole world, existing singly and apart; and therefore, if there be no God, they must needs be absolute non-entities; as immutability, necessary existence, infinity, and perfection; so that the painter that makes this idea, must here feign colours themselves, or create new cogitation and conception out of nothing.

Lastly, if there be no God now existing, it is impossible that ever there should be any, and so the whole idea of God, would be the idea of that, which has no possible entity neither; whereas those other fictitious ideas, made by the mind of man, though they be of such things, as have no actual existence, yet have they all a possible entity.

Cf. ibid. 674.

2. *From the logical axiom* that nothing can come out of nothing. We find a further argument for the existence of God in the axiom, that "nothing can come out of nothing". It has been proved above that ideas cannot be derived from sensible things. From this Cudworth goes one step further and says, if there were no perfect *νοῦς* as the first *νοητόν* (object of knowledge), then the ideas would come out of nothing[1].

3. *From the necessity of Truth.* We have seen that in every act of knowledge an absolute relation system of meaning and purpose as "one Whole" is presupposed. From this also follows the existence of God, for, if the absolute *νοῦς* once IS, then it IS from eternity[2]. It is contrary to the idea of absolute Truth that it could have been created; it is essentially self-existent.

4. *From the finite mind.* In the same way it can be said: if a finite mind exists at all, this is, because it is founded upon an absolute mind, without which there would be neither standard or knowledge. The existence of God is directly implied in the necessity of Truth itself. It follows directly from the fact and nature of finite knowledge and is already presupposed in the fact, that we have within us the idea of God at all[3].

The existence of God follows from Being as well as from knowledge[4], by virtue of the parallelism of Being and knowledge, of life and thought.

5. *From possible Being.* From the point of view of the actual existence of material things, the ideas can be considered as the "Possibilia" of all things. They are possible, because a perfect power ac-

[1] Ibid. 872. Were there no perfect mind, viz. that is an omnipotent being comprehending it self, and all possibilities of things vertually contained in it; all the knowledge and intelligible ideas of our imperfect minds, must needs have sprung from nothing.

[2] Ibid. 734.

[3] Ibid. 729. If once there had been no mind, understanding or knowledge, then could there never have been any mind or understanding produced.

Mor. 254.

[4] T. I. S. 729. If once there had been no life, in the whole universe, but all had been dead, then could there never have been any life or motion in it.

tually exists, who is capable of bringing all possible Being into actual existence[1].

6. *From temporal Being.* Contingent and temporal Being points to a necessary and eternal Being, because, if at one time anything exists at all, it can only exist, because there is one perfect Being who has existence in himself and the power to create contingent life[2].

7. *From the idea of imperfect Being.* The existence of God is directly implicit, not only in the idea of God, but also in the idea of imperfect Being, because only by the standard of the perfect can the imperfect be recognised as imperfect and measured in the degree of its relative perfection[3].

8. *From supernatural events.* Cudworth finds a further argument for the existence of God from the fact of supernatural events, as for instance, ecstacies, which, as he says, would have no object, if there were no God[4]; or visions, miracles and prophecies which, without God, would have no author[5]. He justified himself for believing in

[1] Ibid. 749. ... as for all other things, which are in their own nature, contingently possible, to be or not to be, reason pronouces of them, that they could not exist of themselves necessarily, but were caused by something else.

Ibid. 733. ... knowledge and understanding ... in their nature, do plainly suppose the actual existence of a perfect being, which is infinitely fecund and powerful, and could produce all things possible or conceivable.

[2] Ibid. 738. Nothing which once was not, could ever of it self come into being.

Ibid. 715. All successive things, the world, motion, and time, are in their own nature absolutely uncapable of an ante-eternity, and therefore there must of necessity, be something else of a permanent duration, that was eternal without beginning.

[3] T. I. S. 648. Now, that we have an idea or conception of perfection, or a perfect being, is evident, from the notion that we have of imperfection so familiar to us: perfection being a rule and measure of imperfection.

... perfection is the first conceivable, in order of nature, before imperfection, as light before darkness and positive before privative or defect ... for perfection is not properly the want of imperfection, but imperfection of perfection.

Ibid. 696. Whereas they (the atheists) could not otherwise judge, the greatest perfection and happiness which ever they had experience of in men, to be imperfect, than by an anticipated idea of perfection, and happiness, with which it was in their minds compared.

Cf. Plato, Phaedo, 74*e*.

[4] T. I. S. 640.

[5] Ibid. 700 ff.

such supernatural events by a most chivalrous consideration, saying that, to deny the objective truth of the records of such supernatural phenomena, would be a most rude vote of distrust in the sincerity and honesty of the whole of mankind; a distrust which would not strike only at these records, but at all the records of history. Even more formidable, it would affect sense-perception itself, since the supernatural events are also conveyed by sensations, and thus bring those who deny them into the most awkward situation, because they are commonly those who consider sense perception as the only rightful judge and standard of reality.

These speculations may be of small interest in our present research, but the assumption of higher intelligible beings takes an important place in Cudworth's philosophy. He divides miracles and prophecies into two groups, those which can be performed by higher intellectual beings, and those which can only be wrought by a perfect divine power. It is an idiosyncrasy, very characteristic of Cudworth, that it is in the first group, in which the existence of higher beings is declared, that he finds a greater, indirect proof of the existence of God, than in any argument concerning rational beings, who exist in a vital union with a material body.

We find in Cudworth a noteworthy inconsistency, as regards the relationship between perfect and imperfect Being. On the one hand he repeats again and again that an approach of the finite to the infinite is always an illusion, because a fundamental gulf divides the imperfect from the perfect. This gulf can never be bridged from the side of the imperfect, not by any advance in relative perfection[1]. Before the Absolute in its high transcendency of perfection, any degree of relative perfection must necessarily be meaningless. The gulf between perfect and imperfect can be overcome only from the side of the Perfect, but this inevitably is at the cost of the self-limitation of God.

In the refutation of the atheists, Cudworth emphasises quite a different aspect, because to them the universe, that divine work of art, has become a flat plane, through their deriving life and thought

[1] Ibid. 696. Finite things put together can never make up infinite.
God differs not from these imperfect created things, in degrees only, but in the whole kind.

from matter. To them Cudworth shows the wonderful ladder of Being, perfection and intelligibility in creation. All creation strives ever and ever after the perfection and reality of God; and in him it finds its final consummation[1].

The assumption of higher intelligible beings provides indirectly also a better explanation for the lower kinds of beings, endowed with soul, the animals. Contrary to those who see in the actual differences between higher and lower beings the result of their own use of free will, Cudworth sees the cause of the difference in the different constitution of their soul[2]. He argues as follows: if it were the result of their free will, it would mean that, to each soul, at any time, the whole ladder, from the highest perfection to the lowest deficiency, would be accessible, and that the entire responsibility for its actual state would rest exclusively upon this soul. The soul would then, at every minute, find itself exposed to the most formidable dangers. It would live in a state of inconceivable insecurity and could never enjoy any rest or happiness or peace.

The Demonstration of the existence of God by Descartes[3]

Cudworth quotes the demonstration of Descartes in the following form: the idea of a perfect Being necessarily implies existence, and therefore it IS. He considers it as a renewal of the medieval, ontological demonstration of God.

[1] Ibid. 858. There being plainly a scale or ladder of entity; the order of things was unquestionably, in the way of descent, from higher perfection, downward to lower ... neither are the steps or degrees of this ladder, (either upward or downward) infinite.

A perfect understanding being, is the beginning and head of the scale of entity; from whence things gradually descend downwards; lower and lower, till they end in senseless matter.

Ibid. 648. (Described as ascent.)

Nor indeed, could these gradual ascents, be infinite, or without end; but they must come at last to that which is absolutely perfect, as the top of them all.

(This passage is a good example of how daringly and uncritically Cudworth sometimes expresses himself, even when dealing with a problem which he had critically thought out with the greatest care and circumspection.)

[2] Ibid. 567.

[3] Ibid. 721 ff.

Cudworth objects that the existence of God outside our minds does not follow from the fact that we have the idea of the perfect Being in us. A demonstration of God is not valid or conclusive, if it stands on the accidental indwelling of this idea in our minds.

The "possibility" of the actual existence of God can be inferred from the fact that the idea of God in itself is not contradictory; the actual existence of God can be inferred from the argument that the idea of God itself implies necessary existence, independent of, whether or not, this idea is activated by any finite mind.

Yet from the idea we can never attain to actual existence, and here is the integral difficulty. The only possible conclusion is therefore: if God exists, he exists as necessarily existent Being. But this sentence implies a contradiction, because absolute necessity cannot be inferred from hypothetical premises. But the necessity of existence must be taken in an absolute sense, because the idea of God is necessarily related to existence, in the same way as, contrariwise, the idea of contradiction is necessarily related to the impossibility of existence; or again, as the idea of imperfect Being is related to contingent existence.

Necessary relation to existence means, that a perfect Being can only BE, absolutely and actually, because his existence is a necessary part of his own perfect nature. The hypothetical sentence, "if it is perfect then it is necessary", places the perfect into contingent relations and is therefore incorrect. In this restricted sense we are justified in considering actual existence as being implicit in the idea of the perfect Being itself.

Cudworth thus modifies the Cartesian demonstration upon these considerations in the following way[1]:

a) The existence of God is possible, because the idea of God implies no contradiction.

b) God is necessary, because of the perfection of his nature.

c) From these two premises it follows that God actually exists. For, if God did not exist now, he could not exist at any time, because, as a perfect Being, he cannot have a beginning. A necessarily existent

[1] Ibid. 724 ff.

Being, if it is possible, actually IS, or else it is again thought of as contingent, and not perfect, Being.

In other words, everything which can be thought either IS, or it is possible; but God, if he does not actually exist, is not possible either, which is the same, as to say that he cannot be thought.

Cudworth does not mention the other, more important form, of Descartes' argument, which runs as follows:

The fact, that we know ourselves as imperfect Beings, forces us to suppose a perfect Being as actually existent; for, being ourselves imperfect, the idea of perfection cannot have come from ourselves.

The examination of the form of the Cartesian demonstration of God loses force, if one considers that, as an effort of the accidental discursive reason, it cannot be counted, according to Cudworth, as a metaphysical certainty, capable of being the foundation of knowledge, and of guaranteeing the veracity of our cognoscitive powers. Cudworth himself does not attribute to it any importance or dogmatic weight in his own theory of knowledge[1].

The cosmological demonstration of God

The refutation of Descartes grows almost dramatic, when Cudworth turns to the rejection of the cosmological demonstration of God[2]. He says, that, under a shallow pretence of modesty, Descartes rejected the demonstration of God from the harmony of the universe, saying that we, with our finite minds, cannot claim insight into the counsels of God. Descartes' ultimate purpose was to find sure metaphysical foundations for an independent sober scientific research, untrammelled by schemes of unscientific final causes so apt to obstruct scientific progress through theologically dogmatic considerations.

[1] Ibid. 725, see note 2, p. 82.

[2] T.I.S. 683–4. Wherefore these atomic theists, utterly evacuate that grand argument for a God, taken from the phaenomenon of the artificial frame of things, which hath been so much insisted on in all ages, and which commonly makes the strongest impression of any other, upon the minds of men, they leaving only certain metaphysical arguments for a deity, which though never so good, yet by reason of their subtilty, can do but little execution upon the minds of the generality, and even amongst the learned, do oftentimes beget more of doubtful disputation and

Cudworth is not inclined to acknowledge in this purpose a modest discretion. On the contrary he discloses Descartes' ultimate claim for an autonomous research, founded upon guaranteed cognoscitive powers. Such a purpose must prove an illusion in any sound examination[1]. With the claim of knowledge of the final cause, we make no illegitimate claim of insight into the hidden counsels of God[2].

Cudworth's argument of the final cause runs as follows:

We can comprehend things only through the ideas. The idea is Being, meaning and purpose of things; as such it is their final cause, for which they were created as a part of one divine Order of Being and meaning. Things are good, in so far as they accord with their final cause.

The final cause is not a hidden counsel, shut up in God, but is manifest[3] and has reality in the things themselves, in so far as these ARE

scepticism, than of clear conviction and satisfaction. The atheists in the mean time laughing in their sleeves, and not a little triumphing, to see the cause of theism, thus betrayed by its professed friends and assertors, and the grand argument for the same, totally slurred by them; and so their work done, as it were to their hands, for them.

[1] Mor. 273. ... I conceive it not to be an opinion, but only a certain scheme of modesty and humility, which they thought decorous to take upon themselves, that they might not seem to arrogate too much either to themselves, or to their excellent performances, by not so much as pretending to demonstrate any thing to be absolutely true, but only hypothetically, or upon supposition that our faculties are rightly made.

[2] T. I. S. 685. But the question is not, whether we can always reach to the ends of God Almighty, and know what is absolutely best in every case, and accordingly make conclusions, that therefore the thing is, or ought to be so; but, whether any thing at all, were made by God, for ends and good, otherwise than would of it self have resulted from the fortuitous motions of matter.

Nevertheless we see no reason at all, why it should be thought presumption, or intrusion into the secrets of God Almighty, to affirm, that eyes were made by him for the end of seeing ... and ears for end of hearing ... this being so plain, that nothing but sottish stupidity, or atheistick incredulity, (masked perhaps under a hypocritical veil of humility) can make any doubt thereof.

[3] Ibid. 148. Moreover those atheists, who philosophize after this manner, by resolving all the corporeal phaenomena into fortuitous mechanism, or the necessary and unguided motion of matter, make God to be nothing else in the world, but an idle spectator of the various results of the fortuitous and necessary motions of bodies; and render his wisdom altogether useless and insignificant, as being a thing

and accord with their purpose. The final cause is, therefore, intelligible and essentially reasonable; finally, it points to the one absolute divine order and purpose in all things, to the Truth which is the divine *νοῦς* itself. The things are created according to their Natures, according to their ideas; and they attempt to realise this their purpose and to represent their respective ideas in their actual Being. Or, seen from another angle, they bear the imprint of the divine mind, manifest wherever "one whole", one purposeful relation system, great or small, is presented to the mind. The highest expression of "one whole" is the universe itself, which reveals, in an especially lovely way, that it is created for one single divine order and purpose. We see it in the regular co-operation of all its particular parts into one beautiful harmony, the *εὖ καὶ καλῶς*, beauty and goodness[1].

Yet we must admit that the universe, as one whole, is never given to us as an object of knowledge, and as such it cannot be comprehended, because our knowledge, depending on experience, cannot do more than proceed from one problem to another. This is manifest in the striking fact that we are utterly incapable of judging, whether the world as a whole is good or bad[2]. But we, nevertheless, have, in a certain sense, the anticipation of the universe as one divine order of coherent meaning and purpose[3]. This is possible, because, in every act of knowledge, we participate in the divine mind, who IS the pure act of knowledge and the one divine order of coherent meaning and purpose. But also every particular act of knowledge is knowledge of one whole, of one limited sphere of coherent meaning and purpose, which, firstly, is part of the one unchangeable relation system of meaning and purpose, the *νοῦς* itself; and, as such, it can never be adequately comprehended by a finite mind. Secondly, this limited sphere of meaning is always presented to us as part of a twofold relation system: *a)* it is in a momentary relation with other objects at

wholly enclosed and shut up within his own breast, and not at all acting abroad upon anything without him.

(This passage also shows how alien to a genuine Platonist is the conception of a "Deus absconditus").

[1] Plato, Phileb. 28*d–e*.

[2] T.I.S. 874.

[3] Ibid. 646.

the moment, when we comprehend it; *b)* by the act of knowledge, it is taken up into a specific relation of ideas within us, and thus it is made part of the relation system of ideas, which form our momentary content of consciousness. Both these relation systems, without and within the mind, are accidental; they change from moment to moment, and essentially influence the knowledge of the newly comprehended object, as one whole. In every different relation system, without or within, the object has a differently specified meaning and significance. From this it follows that, in order to know one thing absolutely and perfectly, it is necessary to know the whole universe, with all its particular possible relations, in which it has its place, together with the specific modifications resulting from these. This is only one part, for it would also be necessary for the perfect knowledge of one single object to foresee every possible future connection of ideas in our mind, and the specific modifications resulting from them.

Apart from the knowledge of Being, meaning and purpose of the things, the idea which is their first cause, *αἰτία*, as well as their final cause, there is, for Cudworth, no other possible form of knowledge. The demonstration of God from the harmony, the *εὖ καὶ καλῶς* of the universe, results from an essentially reasonable view, because only reason itself can grasp a relation system of meaning and purpose, a harmony[1].

In the universe the harmonies are more evident than anywhere else. There the imprints of the divine mind can be read, as it were, in capital letters[2]. Yet, whether the object of knowledge is the universe or any limited relation system of meaning and purpose, makes only a difference in degree. The demonstration of God can actually be derived from any knowledge of "one whole" of meaning and purpose, because it finally springs from the nature of knowledge itself, where the spirit comprehends itself in the things and is its own criterion of Truth.

The things ARE essentially their meaning and purpose, and, apart from this, they have no Being. To deny the final cause, therefore,

[1] Ibid. 413.

[2] Ibid. 677. The divine mind and wisdom, hath so printed its seal or signature upon the matter of the whole corporeal world, as that fortune and chance could never possibly have counterfeited the same.

Plato, Phileb. 28*d*.

means no less than to deny the reasonableness of the universe and of all its parts[1]. Since the meaning and purpose is also the creative cause (*αἰτία*) of things, therefore Descartes, by rejecting the final cause, ultimately excludes knowledge of any other cause. Above all, he can never explain any harmony, the *εὖ καὶ καλῶς*, without a final cause, neither in the whole universe nor in its particular parts[2].

The mechanical causes (*συναίτια*), when disconnected from the transcendent *αἰτία*, the divine *νοῦς*, are meaningless, because their significance is found exclusively in their acting towards the realisation of one reasonable system of coherent meaning and purpose, which ultimately is the divine *νοῦς* itself[3].

Every knowledge of ideas is knowledge of the final cause. Since we can only think in ideas, we have no other possibility of knowledge whatever than the knowledge of the final cause. From this point of view, Descartes' aim of scientific research, independent of final causes, proves to be contradictory and a vain illusion[4].

If the final cause is abandoned, any possible transcendental foundation of knowledge falls also, and the universe is given over to mere chance[5]. It makes no difference that Descartes assumes a divine

[1] T. I. S. 669. Or lastly, they deny it to have any cause at all, since they deny an intending cause, and there cannot possibly be any other cause of artificialness and conspiring harmony, than mind and wisdom, councel and contrivance.

[2] Ibid. 175.

[3] Ibid. 672. ... the mechanick powers are not rejected, but taken in, so far as they could comply serviceably with the intellectual model and platform.

Plato, Phaedo, 97*c–d*.

99*b*.

Tim. 69*a*.

[4] T. I. S. 148. ... That all the effects of nature come to pass by material and mechanical necessity, or the meer fortuitous motion of matter, without any guidance or direction, is a thing no less irrational than it is impious and atheistical.

Plato, Phileb. 26*d*.

[5] T. I. S. 415. ... that which acts without respect to good, would not so much be accounted MENS as DEMENTIA, mind, as madness or folly.

Ibid. 147.

Ibid. 54.

Ibid. 676. It is an atheistical, fanatical nonsense to say that arbitrary motion of dead matter could, in time, change into regular methodical action, as if perfect wisdom directed it.

(A gradual transition from chance to law is impossible.)

creator, who has in the beginning transferred to the universe a certain amount of local motion, which, by virtue of his laws, preserves itself and works itself out, but which acts towards no end. Such a creator is for his creation a far removed "terminus a quo", but no longer the "terminus ad quem"; and this contradicts the conception of God.

The essential place of every creature is between the Absolute as its creative cause (αἰτία) and the Absolute as its final cause, so that it is encompassed all round by the Absolute[1]. If God looked upon his creation as something outside and alien to himself, if, as a mere spectator, he left his universe as a perfect machine to run its own course[2], he would, by this confrontation, make himself finite and relative in respect to that with which he confronts himself, and would therefore forfeit his own divinity[3]. A God who is nothing more, than the starting point of his universe, is not a God at all, and it is from such a view a dangerously small step to the denying of the existence of God altogether and believing in the autonomy of matter. Descartes' modest renunciation of the knowledge of the final cause is thus disclosed as "superbia", as a kind of camouflaged atheism. His attitude appears to Cudworth altogether contradictory. The rejection of the final cause implies the denial of the divine creator, and excludes laws as given by God, for these would necessarily be reasonable and directed towards a final cause. Cudworth therefore justly reproaches Descartes for, in fact, deriving the universe from vortices of atoms; and says that the derivation of motion from a divine force is but a sham.

Descartes' own ambiguity allows Cudworth to interpret him in the opposite way also. Thus he says that Descartes, with his assertion of a divine creator, implicitly professes God as the end of all Being and knowledge, and that his rejection of the final cause was only formal.

[1] Cf. T.I.S. 647. ... this being comprehends the differences of past, present, and future, or the successive priority and posteriority of all temporary things.

[2] T.I.S. 683. ... God in the mean time standing by, only as an idle spectator, of this LUSUS ATOMORUM, this sportful dance of atoms, and of the various results thereof. Nay these mechanick theists, have here quite outstripped and outdone, the atomick atheists themselves, they being much more immodest and extravagant, than ever those were. (Because they believe that the universe turned out well at the first attempt.)

[3] Ibid. 382.

From this angle, Cudworth proves that Descartes' conception of natural laws is in truth nothing less than Plastic Nature[1], that is, an immediate efficient cause, created by God, and, though acting unconsciously, directed towards an end, since laws of nature, qua laws, are necessarily reasonable, and this means that they are determined for a purpose.

To assume that Descartes' rejection of the final cause was only meant as a preliminary hypothetical method of scientific research, and that it was not given general validity, does not resolve Descartes' ambiguity, because all true knowledge is precisely knowledge of the final cause, which is the only reasonable cause for the particular things, as well as for the universe. To deny it is to deny the intelligibility of the whole world, for the alternative to "causa finalis" is chance, and chance is essentially unintelligible.

The main purpose in his argument for the validity of the cosmological demonstration of God is for Cudworth not primarily the demonstration of the existence of God. As an accidental act of discursive reasoning it could have no decisive significance for knowledge. Cudworth's ultimate end here also is to prove the transcendental foundation of all finite Being and knowledge in the divine *νοῦς*.

Cudworth sees a still more formidable danger arising from Descartes' assumption that Truth itself was created by an arbitrary act of God, for, with it, he hands over to chance, not only the universe, but Truth itself, and so destroys the foundations of knowledge altogether. This leads us on to Cudworth's heaviest attack, which strikes at the whole field of possible knowledge, logic, ethics, and theology; and therewith we come to his exposition of the attributes of God.

[1] Ibid. 151. ... and that their laws of nature concerning motion, are really nothing else, but a plastick nature, acting upon the matter of the whole corporeal universe, both maintaining the same quantity of motion always in it, and also dispensing it, according to such laws, fatally imprest upon it.

Part II

THE ATTRIBUTES OF GOD; THE DOCTRINE OF THE TRINITY

All we are able to say of God is summed up in the one attribute, Perfection[1]. The perfect cannot be comprehended by us in an adequate way. All our assertions are but an attempt to symbolise, in finite form, that which cannot be uttered.

We can symbolise the Perfect as *πέρας* and as *ἄπειρον*[2].

The Perfect is *πέρας*, because it is eternally identical with itself, unchanging in the face of the tendency of all finite Being to disintegrate, from moment to moment, into a purposeless infinity. As limit, the Perfect bestows Being and purpose to that which is becoming. It is its norm and paradigma.

The paradoxical sentence that the ladder from Matter up to God cannot be infinite may here be found explicable. Created Being, although it has in itself the tendency to disintegrate into a purposeless infinity, can never quite lose itself, because it depends on the ideas. The 'Possibilia' of things are restricted and limited by the immutability of their nature, and by the fact that the divine mind, just because it is perfect and infinite, cannot transcend its own content of ideas. Though this is a paradox, it fitly symbolises the perfection of the *νοῦς* as infinity, yet not infinity in the sense of a shapeless indefinite abyss.

The Perfect is *ἄπειρον* when it confronts the limitation, inherent in the finite, whose mode of duration is succession, and which, passing from present moment to present moment, losing the past with every new present, can always only be one part of itself, and never itself as a whole. The infinity of the Perfect can be seen precisely in the fact that it keeps its identity eternally, never losing anything of itself, never particularising into indefinite parts.

The true meaning of infinity is Perfection[3]. There is, therefore, neither infinite number, infinite space, nor infinite time. For to all num-

[1] Ibid. 200. The true and proper idea of God in its most contracted form is this: a being absolutely perfect.

[2] Ibid. 389–90.

[3] Ibid. 647. Infinity is really nothing else but perfection.

ber, space, and time something can yet be added. But, in their faculty of indefinite increase, they imitate infinity in a finite way[1]. Since number, space, and time can never be perfect, they cannot be attributed to a perfect Being[2], and when we speak of God as the transcendent ἕν (One) we do not mean the figure one, but one as the principle of the scale of numbers[3].

Since our thought and imaginations are activated in the forms of space and time, the Perfect as such can never enter either our imagination or thought. Space is the mode of being of the finite, it could never receive or stand up to the direct assault of the absolute creative power[4].

But a different relationship exists between the Perfect and time. Everything has its own mode of duration. Eternity is the mode of duration of the infinite[5], which IS ever identical with itself in the eternal actual present. Eternity is not a positive attribute, but a mode

[1] Ibid. 644 ff. See Note 3, p. 3.

[2] Ibid. 409. (All things are dependent upon God, therefore there is neither space, number nor time in him.)

Ibid. 647. Infinity of duration or eternity, is really nothing else, but perfection, as including necessary existence and immutability in it. So that it is not only contradictious to such a being, to cease to be, or exist; but also to have had any flux or change therein, by dying to the present, and acquiring something new to it self which was not before.

[3] Ibid. 373.

[4] Ibid. 775. And if there were corporeal empty space, then reason, who is God could not exist in it as coextended with the same, for space can only receive body, it could not receive the power of the intellect, nor could the intellect give to space that which is not corporeal.

(Cudworth acknowledges, according to his definition of the creature "as that which has a beginning in time", that God can act ad extra only with time. But the work of God is not restricted to space, for he can also act cogitatively upon cogitative beings.

Moreover, Cudworth leaves the question open whether all creatures must necessarily have some kind of a body as vehicle.

The essential and genuine work of God is that within the Trinity. This shows how far Cudworth is from Newton's conception as absolute space of SENSORIUM DEI, and how clearly he also here keeps to the Platonic line.)

[5] T. I. S. 388. The supreme deity ... whose duration therefore was very different from that of ours, and not in a way of flux or temporary succession, but a constant eternity, without either past or future.

"A standing eternity.".

of being[1]. The duration of the finite is time, because it has a beginning and can realise itself successively only. Even God, when He directs His creative power "ad extra", can create with time only, because the finite cannot exist in the mode of eternity. It means that God can create finite Being only, when He restricts His own absolute power, or the finite Being would be crushed under its force[2]. In this sense time is also in God.

The Becoming of the things towards their final cause is inseparable from time, since a purpose can only be set in the future. Time is thus linked to the final cause, and thereby to the divine mind itself, who is the ultimate end for all created Being[3].

If there were infinite and perfect time, we could never arrive at a beginning[4], but should be caught in an indefinite regression. From

[1] Ibid. 645. The atheists here, can only smile or make faces; and show their little wit, in quibbling upon NUNC STANS, or a standing Now of eternity; as if that standing eternity of the deity were nothing but a pitiful small moment of time standing still; and as if the duration of all beings whatsoever must needs be like our own. Whereas the duration of every thing, must of necessity be agreeable to its nature; and therefore, as that whose imperfect nature is ever flowing like a river, and consists in continual motion and changes one after another, must needs have accordingly a successive and flowing duration, sliding perpetually from present into past, and always posting on towards the future, expecting something of it self, which is not yet in being, but to come.

So must that, whose perfect nature, is essentially immutable, and always the same, and necessarily existent, have a permanent duration; never losing any thing of it self once present, as sliding away from it; nor yet running forward to meet something of it self before, which is not yet in being: and it is as contradictious for it, ever to have begun, as ever to cease to be.

[2] Ibid. 887. Now we say, that the reason why the world was not made from eternity, was not from any defect of goodness in the divine will, but because there is an absolute impossibility in the thing it self; or because the necessity and incapacity of such an imperfect being hindered. For we must confess, that for our parts, we are prone to believe, that could the world have been from eternity, it should certainly have been so.

[3] Plato, Tim. 39*e*.

[4] T. I. S. 644. Lastly, we affirm likely concerning time or successive duration, that there can be no infinity of that neither, no temporal eternity without beginning: and that not only because there would then be a natural infinity and more than an infinity of number; but also because upon this supposition, there would always have been an infinity of time past. ... which was never present. Whereas all the moments of past time, must needs have been once present; and if so, then all of them, at least,

each moment of time there would necessarily have to stretch an infinity of past time, which would never have been present. The conception of time itself implies that one day was the first.

Further, it would also have to be possible to contract and comprehend an infinity of future time into one present.

Through the specific relationship of time to the creator and to the final cause, time also stands in a special connection with eternity. Time and eternity cannot be put in contrast to each other, as if time were outside of eternity; neither can time be explained merely subjectively, as the form of imagination of finite rational beings. Time is 'in eternity', it is its image, and, apart from eternity, it cannot be thought, nor can it be disconnected from eternity[1].

Since the Perfect can be symbolised both as *πέρας* and *ἄπειρον*, it can also be considered as neither of them, but as beyond any possible predication, as absolute Oneness; not one, however, in contrast to the many, but the pure transcendent One.

Cudworth's exposition of the attributes of God

It is not an easy task to give a clear account of Cudworth's exposition of the attributes of God, since the critical and dogmatical strands of his thought are here almost inextricably worked together.

All the attributes of God are inadequate conceptions; they have significance only *πρὸς ἡμᾶς* (for us). They are an unfolding in finite thought of the one idea of the Perfect[2] into self-existence, necessary

save one, future too; from whence it will follow, that there was a first moment or beginning of time. And thus does reason conclude, neither the world nor time it self, to have been infinite in their past duration, or eternal without beginning.

[1] T.I.S. 572. ... all temporary beings, once to have had a beginning of their duration...

... time it self was not eternal.

Plato, Tim. 37*d*.

Tim. 38*b*.

T.I.S. 396. (Time is an image of the uncreated duration of eternity. As the sensible world was created as an image of the intelligible world, so was time created together with the world as an image of eternity.)

[2] Ibid. 652. In very truth, all the several attributes of the deity, are nothing else but so many partial and inadequate conceptions, of one and the same, simple per-

existence, eternity, infinite goodness, wisdom and power. The persons of the Trinity exist independently of this attempt of finite minds to comprehend God, by means of his attributes[1].

In spite of this critical restriction of the *πρὸς ἡμᾶς*, Cudworth asserts that the attributes of God constitute the idea of the Perfect *φύσει* (by nature) and are independent of their being thought by finite minds. They are necessary and demonstrable, like geometrical theorems; they are all known to us, because they form one, necessary, simple Whole[2]. The inadequacy of our comprehension of them lies merely in this, that we are not able to grasp fully their meaning and purport[3], just as we can, for instance, know a geometrical figure without the explicit knowledge of its definition.

This assertion, however, contradicts Cudworth's critical exposition of the theory of knowledge, where he says that our knowledge, incited by experience, proceeds successively, so that we comprehend Truth always only in a very small section, because it is bigger than we are, and we are not identical with it. Even if we filled ourselves with the knowledge of God to the full, God would still be as the ocean, and we as little particles lost in it. How then could we, through discursive reasoning, attain to, at least formally, an adequate comprehension of the Perfect?

Yet from another aspect these startling assertions might become somewhat more explicable. Cudworth oftens speaks of the demonstrability of the divine attributes. He obviously means this in the sense that the conception of God belongs to the sphere of pure and necessary

fect being, taken in as it were by piece-meal: by reason of the imperfection of our humane understanding, which could not fully conceive it altogether at once.

And therefore are they really all but one thing, though they have the appearance of multiplicity to us.

[1] Ibid. 558. It is not a trinity of meer names or words, nor a trinity of partial notions and inadequate conceptions, of one and the same thing.

[2] Ibid. 652. ... all of them attributes of nature, and of most severe philosophick truth.

But all the genuine attributes of the deity, of which its entire idea is made up, are things as demonstrable of a perfect being, as the properties of a triangle or a square are of those ideas respectively.

[3] Ibid. 653. So neither doth every one, who has a conception of a perfect being, therefore presently know all that is included in that idea.

ideas, where, as far as the object of knowledge is concerned, adequate knowledge and evidence can be attained. God is the most intelligible Being[1]. The reason for the inadequacy of our knowledge of God is the abundance of pure Truth in Him, and not His being contrary to Reason.

In our highest knowledge we reach only to the pure ideas in logic and mathematics; but God, as the absolute One Perfect, *ἓν ἀγαθόν*, is higher than ideas. Yet we cannot speak or think of God, except through ideas. Since, however, evidence can be obtained within the sphere of pure ideas, and God, the most intelligible, comes forth from His absolute unutterable Oneness into the multiplicity of the ideas, no element has, as yet, come in which could darken the evidence. The reason that we do not actually attain evidence in the knowledge of God lies exclusively in the manner in which the ideas are remembered in us, namely that they are in so many ways exposed to chance.

Cudworth's assertion that the attributes of God are necessary and complete can be supported also by the argument that they are all merely an unfolding of the one idea of perfection and already actually included in it[2]. Therefore, all the attributes presuppose the singleness of God[3]. The anticipation of a perfect Being is never found apart from a genuine monotheism.

Yet a further difficulty arises from Cudworth's assertion that all the attributes belong to the Trinity as a whole, but, at the same time, to each Person also; whereby the attributes, Goodness, Wisdom, and Power, which are on the one hand considered as part of our inadequate conception of the Perfect[4], represent on the other hand the essence

[1] Ibid. 639. It is true indeed, that the deity is more incomprehensible to us than any thing else whatsoever, which proceeds from the fulness of its being and perfection, and from the transcendency of its brightness, but for the very same reason, may it be said also, in some sense, that it is more knowable and conceivable than any thing.

[2] Cf. ibid. 636. (The idea of God is the purest idea.)

[3] Ibid. 210.

[4] Ibid. 558. Concerning the Christian Trinity, we shall here observe only three things, first that it is not a trinity of meer names or words, nor a trinity of partial notions and inadequate conceptions, of one and the same thing. For such a kind of trinity as this, might be conceived, in that first Platonick hypostasis it self, called *τὸ ἓν ἀγαθόν*, the One and the Good, and perhaps also in that first Person of the Christian Trinity; namely of Goodness, and understanding or Wisdom, and Will or active power, three inadequate conceptions thereof.

of each particular Person[1] to such a degree that they are almost identical with them. Thus the whole of Cudworth's doctrine of the Trinity remains in a strange ambiguity[2].

Cudworth's doctrine of the Trinity

In a long digression on the history of religion and philosophy, Cudworth proves that the anticipation of the divine Trinity is universally present in the religions and philosophies of all ages. He sees in this a sign of the profound reasonableness of the conception of the Trinity[3]. Yet to prove this is not his chief aim; beyond it he endeavours to show that the Trinity is Knowledge and Truth itself; and thus in the Trinity logic, ethics, and religion have their ultimate foundations.

Cudworth distinguishes two functions of the Trinity, Its activity 'ad extra' and 'ad intra'.

The Trinity reveals Itself 'ad extra' as Will, as self-existence, perfect *αἰτία* (creative cause) for the creation and preservation of the world and the progression of its life and history.

Cudworth symbolises the life of the Trinity 'ad intra' as 'continuity without a Between', *μεταξύ*, as perfect inexistence of one within the other, a relation which does not imply mutual relativation of the parts, such as is never found among creatures[4]. Nor can this relation

[1] T. I. S. 600.

[2] Ibid. 587. For if the whole Deity, were nothing but one simple monad, devoid of all manner of multiplicity; as God is frequently represented to be, then could it not well be conceived by us mortals, how it should contain the distinct ideas of all things within it self, and that multiform platform and paradigm of the created universe, commonly called the archetypal world.

[3] Ibid. 560. We shall conclude here with confidence, that the Christian Trinity, though there be very much of mystery in it, yet is there nothing at all of plain contradiction to the undoubted principles of humane reason, that is of impossibility to be found therein.

[4] Ibid. 559.

Ibid. 597 ff. (The persons of the Trinity.) First, because they are indivisibly conjoyned together, as the splendour is indivisible from the sun. And then because they are mutually inexistent in each other, the first being in the second and both first and second in the third. And lastly because the entireness of the whole Divinity, is

be fully intelligible to any finite mind. The fact, however, that substances, so different as Cogitation and Extension, can yet enter a relationship of intimate vital union, might point to yet higher possibilities of unity in multiplicity, beyond the explicit grasp of finite reason[1].

For the relationship between the Persons of the Trinity, Cudworth seeks a 'via media' between the doctrine of *μονοουσία* and that of Tritheism[2].

His argument against Monoousia is as follows: if the Persons of the Trinity are considered merely as different manifestations of the one God, then any inter-relation or action between the Persons is impossible. In consequence, the finite world remains the only field of action for the creative power of God. God is thereby brought into a certain dependence on his creation, since He can work only in the realm of the finite and act upon an object necessarily incongruous to Him[3]. The eternity of God would then demand, as a necessary consequence, the eternity of the world; yet we have seen that the finite cannot exist in the mode of eternity.

The assumption of Tritheism leads into no less a difficulty. Since oneness is implicit in the conception of perfection itself, it is contradictory and irrational to conceive the Trinity as three Persons who are absolutely co-ordinate and existing 'without' each other. The Persons would necessarily be relative to one another, and their being three would also be arbitrary; if once the paradox of a plurality of self-existent beings is admitted, their number cannot reasonably be restrict-

made up of all these three together, which have all *μίαν ἐνέϱγειαν*, one and the same energy or action AD EXTRA.

Ibid. 617 (erroneously numbered 601). That these are so nearly and intimately conjoyned together, that there is a kind of *συνέχεια* continuity betwixt them. ... a mutual inexistence in one another.

[1] Ibid. 559. We grant indeed, that there can be no instance of the like unity and oneness found in any created beings; nevertheless we certainly know from our very selves, that it is not impossible, for two distinct substances, that are of a very different kind from one another, the one incorporeal, the other corporeal, to be so closely united together, as to become one animal and person; much less therefore should it be thought impossible, for these three divine hypostases, to be one God.

[2] Ibid. 612.

[3] Ibid. 546.

ed or fixed. The reason for the Triad can only be sought in the mode of their inter-relation itself.

Cudworth comes to the conclusion that the Trinity cannot be comprehended at all, without the assumption of a certain subordination 'ad intra'[1]. Without any kind of difference, there could be neither inter-relation nor action between the Persons[2]. The belief in a Trinity would then become entirely superfluous and meaningless. Subordination, however, can be assumed only in their intimate relationship 'ad intra', and within the absolute oneness of their mutual inexistence.

Cudworth refutes three Neo-Platonic errors, concerning the Trinity.

1. The progression from the ἓν to animal souls. Cudworth rejects the Neo-Platonic view of the Trinity as a progressive dividing out of the transcendent One into multiplicity, whether this process be understood as taking place in three strides or as a continuum. He, nevertheless, appreciates the desire lying behind this view, to preserve the strict oneness of the transcendent cause beyond Being and Knowledge. Yet, from the conception of such a progressive division, it is obviously but a small step towards considering the third 'hypostasis' ψυχή as World-Soul, which is immersed in matter and directs the universe in vital union from within, in analogy to the finite soul and its particular body[3]. If God were, as World-Soul, vitally united to the universe, He would be exposed to passive suffering by His necessary union with it, and thus made relative. The finite world would then be part of the Trinity, and the fundamental difference between creator and creature effaced, and the Triad become altogether arbitrary.

In vital union as such, the frontier between spirit and matter is not necessarily passed. We have seen that, in spite of the assumption of vital union, Cudworth asserts the fundamental difference of human souls from their bodies. Yet vital union between the Perfect and matter is impossible, because it naturally implies a reciprocal influence between the united parts, and this is totally inconsistent with the Perfect, which is pure activity. God, conceived as soul, can only be creator and

[1] Ibid. 598. (Subordination AD INTRA according to rank and dignity.)
Ibid. 611. (No ταυτοουσία.)

[2] Ibid. 607. (It is just the "equality of nature" which implies a certain unlikeness of the hypostases.)

[3] Ibid. 552.

preserver of the world in perfect independence from his creatures as ψυχὴ ὑπερκόσμιος[1]. His decision to create the finite universe is contingent, not necessary; it is not part of the nature of God. Whenever Cudworth considers Plastic Nature as World-Soul, immersed in matter and directing it from within[2], he always emphasises its creatureliness, finiteness, and complete dependence on God[3].

2. A further error in the Neo-Platonic conception of the Trinity is that the *νοῦς* as object of knowledge, the ideas, is severed from the *νοῦς* as act of thinking. Thus the content of knowledge is irrationally dissolved into a multiplicity of independent self-existent 'Substantiae Separatae', which eventually were deified[4]. Thereby the unity of the divine mind is broken up and the Trinity dissolved.

3. The third error consists according to Cudworth in a linear subordination of the Hypostases. In a gradual descent the transcendent One sends itself out into the Many; the divine is conceived as a divine sphere, which can be symbolised as a continuum, upon which any number of points can, ad libidum, be fixed; the result is the interpolation of intermediary Hypostases; thus *ἑνάδες* ('Ones') were interpolated between the *ἕν* and the *νοῦς; νόες* between the *νοῦς* and the *ψυχή*; and *ψυχαί* from the *ψυχή* down to animal souls. They represent a continuous descent from the transcendent One down to the animal souls[5]. The fixation of three main hypostases on this continuum appears as arbitrary and unjustified.

The consequence of such a conception is fatal. The *ἑνάδες* are particular, not universal; they are relative, creaturely, and not perfect, since they are not one. The *νοῦς*, in its strict succession, is made relative

[1] Ibid. 561.

[2] Ibid. 156. But as God is inward to every thing, so nature acts immediately upon the matter, as an inward and living soul or law in it.

[3] Ibid. 155*b*. Nature is "Ratio mersa" et "confusa", reason immersed and plunged and as it were fuddled in it, and confounded with it. Nature is not the divine art archetypal, but only ectypal, it is a living stamp or signature of the divine wisdom; which though it act exactly according to its archetype, yet it doth not at all comprehend nor understand the reason of what it self doth.

[4] Ibid. 563.

[5] Ibid. 555 f.

556 ... a gradual descent of things from the first or highest, by steps downward, lower and lower, so far as to the souls of all animals.

and finite by them, and the PSYCHE suffers the same relativation by both, *ἑνάδες* and *νόες*. Thus, ultimately only the first hypostasis, the One, remains in divine perfection, and even this is questionable, since it is the starting point of the continuum, which continues unbrokenly down to created Being; in so far as the One is itself part of this continuum, it necessarily becomes relative also.

The divine is here mixed with created Being[1] also through the fact that the continuum reaches down to the *ψυχαί* (animal souls), which, by their vital union with matter, are exposed to passive suffering from the material world. Once the gulf between the divine and finite corporeal substance is passed, there is no reason for maintaining the fundamental difference between corporeal and incorporeal substances either.

Cudworth symbolises the relationship between the Persons of the Trinity, both by a straight line and by a circle. The two aspects do not exclude, but complement, each other. When we attempt, as we do here, to predicate the divine, we cannot expect to achieve more, than to find various expressions which, from different aspects, hint at that which can never be uttered in human words.

If the relationship between the three Persons of the Trinity is symbolised as a straight line succession, the second Person is assumed to proceed from the first Person, and the third Person from the first and the second, as the Creed of the Western tradition indicates in the 'filioque'. Generation in the Trinity signifies a co-eternal emanation and communication of the whole being from Begetter to Begotten, such as can never be found in the procreation of finite beings[2]. It is totally different from any willed creative act. If the relationship between the Persons is symbolised as a circle[3], the emphasis lies in the inexistence of the Persons in each other, where the Holy Ghost is

[1] T. I. S. 557. But as it were melting the Deity, by degrees, and bringing it down lower and lower, they made the juncture and commissure betwixt God and the creature, so smooth and close, that where they indeed parted, was altogether undiscernible.

[2] Ibid. 587. Because this is but an imperfect generation, where that which is begotten, doth not receive its whole being originally from that which did beget, from God and nature.

[3] Ibid. 601.
Cf. Plato, Phileb. 30*c–e*.
Tim. 30*b*.

worshipped as the love with which the Father loves the Son and the love of the Son for the Father.

The Trinity as foundation of knowledge

All certainty of Truth and all possibility of knowledge depends, for Cudworth, on the question, whether Truth itself is eternal and self-existent, or whether it is created; in theological terms, whether the *λόγος* was created by an act of will at a certain time, or whether, as a Person of the Trinity, it is universal, eternal and divine. It is the ancient question of Arius: *ἦν ποτε ὅτε οὐκ ἦν* (was there a time, when he was not ?), upon which, according to Cudworth, hangs, not only the Christian faith, but also all knowledge[1]. Cudworth carefully examines, whether we are justified in applying the Platonic hypostases *ἕν*, *νοῦς*, *ψυχή* to the Persons of the Trinity[2]. He himself is inclined to do so, and interprets them by the attributes of Goodness, Wisdom and Power; or Love, LOGOS and Life. Cudworth was grievously misrepresented in this, and chiefly attacked for the assumption of a certain subordination in the relationship of the divine Persons[3].

This 'subordination' 'ad intra', far from being merely adopted from the Neo-Platonic scheme, is the central point in Cudworth's philosophy, where his own original thought most freely communicates itself to us. Not only all Truth, but also all morality and religion, and above all, the Christian faith, hang precisely upon this 'subordination'. Cudworth's exposition of it is extremely beautiful. The Trinity is indeed the very pulse of his philosophy, the focal point, where all his thought converges. It is the one source and the one end for logic, ethics, and religion. That these are conceived as coming from one common source, and moving to one end, is the secret power of true Platonism and Christianity alike, which makes it possible for Platonism to be the sure foundation of a Christian philosophy. This

[1] Ibid. 575. To conclude, no Platonist in the world, ever denied the eternity of that *νοῦς* or universal mind, which is the second hypostasis of their trinity.

[2] Plato, Phileb. 26*e*.

[3] T. I. S. 592. (Cudworth here emphatically denies the Neoplatonic subordination within the trinity.)

never ceased to move Cudworth to the deepest joy in and praise of the mysteries of divine Providence[1].

The decisive factor in the comprehension of the Trinity is the interpretation of the omnipotence of God.

The Calvinists saw the glory of God chiefly in his power, which they interpreted as an unrestricted, arbitrary will which creates both, good and evil, alike, and is guided by no goodness and wisdom. Omnipotence as such is the goodness and wisdom of God, regardless of whether it works evil or good[2]. Cudworth thinks that Descartes consistently developed this same conception of divine omnipotence, when he subjected Truth itself to its arbitrary creative action, asserting that logical and mathematical axioms were established by an arbitrary act of God once for all. However, once created, they have absolute validity even for God himself, because the divine will, which created them, is eternal and immutable[3]. Therefore even an arbitrarily created Truth may still carry absolute certainty. Descartes thinks that divine omnipotence would be restricted and relativated, were it regarded in its creative power as subject to the rule of eternal, intelligible "possibilia", or Natures. Thus it would depend upon something outside God. He obviously does not think in Trinitarian terms, therefore he cannot reckon with a relationship and action within God, of goodness, wisdom, and power, as Cudworth does. Yet his demonstration of God shows that he, contrary to Calvin, interprets divine perfection as absolute goodness, and identifies the divine goodness with divine power, when he says that it would be contrary to the perfect nature of the creator to have committed his creatures to inevitable deception. Cudworth conceives the problem in an entirely different way. He shows that, if Truth is considered as created, it is inevitably made finite for the sake of the absoluteness of divine power, but that thereby, not only the founda-

[1] Ibid. 594.

[2] Fr. W. 84. Wherefore the matter can be resolved into nothing else but ... God's absolute power, and his arbitrary and unaccountable will, which by reason of his omnipotence makes that to be just whatsoever he will do.

[3] T. I. S. 646. This being plainly to destroy the Deity, by making one attribute thereof, to devour and swallow up another; infinite will and power, infinite understanding and wisdom. For to suppose God to understand and to be wise only by will, is all one as to suppose him, to have really no understanding at all.

tions of knowledge are destroyed[1], but divine power, far from being absolute, is made relative also, because, as "absolute arbitrariness", it is accidental and subject to chance.

Infinite power is perfect power, and the only standard for perfection is the divine *νοῦς* itself[2]. Perfect power is, therefore, power which is guided by wisdom; it is a will which wills only that which has meaning and purpose[3], a creative power which gives form and life to Being, which, in the divine *νοῦς*, is from eternity realised as one divine Order of meaning and purpose. God, when He creates, does not clothe with existence independent essences, as if there were something lacking to the Being of the ideas. The ideas are Being which is realised in perfection in the divine *νοῦς*, and, because of their pure intelligibility, they have more Reality than created things[4]. It would be a paradox, if they had to be clothed with existence by a creative act, in order to attain to their full Reality.

[1] T. I. S. 646. Renatus Cartesius, though otherwise an acute philosopher, was here no less childish, in affirming, that all things whatsoever, even the natures of good and evil, and all truth and falsehood do so depend upon the arbitrary will and power of God... He only adding, that ... when they were once settled by the divine decree, became immutable, that is I suppose, not in themselves or to God, but unto us.

Than which, no paradox of any old philosopher, was ever more absurd and irrational; and certainly, if any one did desire, to persuade the world, that Cartesius, notwithstanding all his pretences to demonstrate a Deity, was indeed but a hypocritical theist, or personated and disguised atheist, he could not have a fairer pretence for it out of all his writings, than from hence.

Fr. W. 50. But this hypothesis of Cartesius alike overthrows all morality and science at once, making truth and falsehood as well as the moral differences of good and evil meer arbitrary things, will and not nature; It thereby also destroys all faith and trust or confidence in God, as well as the certainty of christian religion.

[2] T. I. S. 719. Conception and knowledge are hereby made to be the measure of all power; even omnipotence or infinite power it self being determined thereby; from whence it follows, that power hath no dominion over understanding, truth and knowledge, nor can infinite power make any thing whatsoever to be clearly conceivable. For could it make contradictious things clearly conceivable, then would it self be able to do them; because whatsoever can be clearly conceived by any, may unquestionably be done by infinite power.

[3] Ibid. 647. Infinite power is nothing else but perfect power, that which hath no defect or mixture of impotency in it.

[4] (Cudworth does not keep to this view quite consistently. He occasionally also speaks of ideas as *only* possible being vis-a-vis which ideal being, brought into existence by a creative act of God, has a higher reality.)

In the creative act the sensible things are created together with their "causa formalis", according to their Natures, their respective ideas in the divine νοῦς[1]; which latter always remain transcendent. As finite things they have a beginning, and therefore Becoming in themselves; they cannot BE so perfectly, as the ideas ARE, but they are "becoming towards Being", towards the ideas, in so far as finite and temporal Being is capable of imitating the eternal. All that which exists in the sensible world has Being, meaning and purpose from the ideas[2].

(Cf. Plato's conception of the creation of the Universe, the demiurge as PSYCHE, creates the universe by looking up to the ideas[3].)

It is therefore impossible to conceive the "causa formalis" as capable of being detached from sensible things. A thing instantly ceases to be when it falls out of the relationship with its idea, thus losing its meaning and purpose, and therewith its proportion.

The double significance of the Essences, mentioned above, as transcendent and immanent, finds here a further precision through the distinction of the transcendent idea from the immanent "causa formalis".

A thing, by being created according to its immutable Nature, IS what it IS, and is enabled to preserve, though in a somewhat hazardous way, its own identity in the flux of becoming; only in so far as it achieves this, can it be comprehended and known.

Thus the possibility, both of existence and knowledge of things, rests upon the actual existence of the perfect νοῦς as pure act and content of knowledge. Yet, on the other hand, the ideas, in themselves independent of the Being or Not-being of contingent things, created after their image, are the "Possibilia" for the sensible things, only because there is an actual will with power to create possible Being in sensible forms[4].

[1] Mor. 14. It is universally true, that things are what they are, not by will but by nature.

[2] Plato, Tim. 50*c*.

[3] Plato, Tim. 30*a–b*.

[4] T.I.S. 732. But when from our conceptions, we conclude of some things, that though they are not, yet they are possible to be; since nothing that IS not, can be possible to be, unless there be something actually in being, which hath sufficient power to produce it; we do implicitly suppose, the existence of a God or omnipotent

To say that God creates things according to their Natures is to say that He creates them without violence or force, but, as it were, from within, gently leading them into existence, enforcing on them no alien natural law from without. Such a verbal law matter could not understand. This is, for Cudworth, a further proof of the goodness of the world [1], and of the perfection of its creator, that this creation according to Nature is not counteracted by any resistance on the part of the material. This "from within" and "according to nature" is simply a different expression for, and aspect of, Plastic Nature.

It is interesting to see, how the dogmatic and critical strands are here strangely worked together. On the one hand Cudworth contrasts matter to a verbal law; on the other hand he asserts that there is no Being whatever, neither any existing matter, apart from "law", which is the proportion of all things.

The "Nature" of the things is their total Being. With the conception of a law given to matter from within, Cudworth refutes the dualism of substances in the Cartesian scheme, and if this "action from within", attributed to Plastic Nature, may be understood as a critical attempt to overcome the dualism of substances by the idea, we understand better, why Cudworth put so much emphasis on the conception of Plastic Nature; although, within the theory of substances, it does not, despite the vital union, solve the crucial problem of interaction between the fundamentally disparate substances.

If the Natures are created by an act of God, they are necessarily mortal; for all created Being lives in the flux of Becoming, incapable

being thereby, which can make whatsoever is conceivable, though it yet be not, to exist.

[1] Fr. W. 77. God being infinite disinterested love displaying itself wisely, therefore producing from its fecundity all things that could be made and were fit to be made, suffering them to act according to their own natures.

Ibid. 77–8. And since all rational creatures have essentially this property of LIBERUM ARBITRIUM ... belonging to them, to suppose that God Almighty could not govern the world without offering a constant violence to it, never suffering them to act according to their own nature, is very absurd.

Cf. T. I. S. 886. (Cudworth's joy of the world.)

But the atheist demands, what hurt had it been for us, never to have been made? And the answer is easie, we should then never have enjoyed any good; or been capable of happiness.

Plato, Tim. 48*a*.

of preserving its identity unchangingly, because it is, by nature, unfit to live in the mode of eternity. And if, by a miracle, eternity could be bestowed upon them, yet, even then, immortality would not be theirs in truth; it would always remain but an accidental gift of the creator [1]. That which once was created can, at any time, be annihilated by the same will which created it. It can never be absolute. Ideas, if they were accidental, and not in themselves identical, necessary and immutable, could never constitute any absolute and universal Order of the Whole. Together with absolute Truth, we inevitably lose the transcendental foundation of knowledge, and, with it, every possibility of knowledge altogether [2].

The measure of power is the idea [3]. A will, which does not act in accordance with the *νοῦς* has neither continuity nor reasonable and purposeful intention. It reveals itself as a blind, irrational impetus [4], which, by the arbitrariness of its action, makes itself finite, until finally it dissolves into impotence [5]. Arbitrariness destroys the very essence of will and of power. If creating arbitrarily, God would forfeit His own perfection and divinity, and not be God at all.

To say, that the divine will acts according to ideas, simply means that it is guided by the perfection of its own Nature; and to say, that

[1] T. I. S. 194. ... If it were supposed to have happened by chance to exist from eternity, then it might as well happen again to cease to be.

[2] T. I. S. 717. ... So long as they adhere to that perswasion; they can never be assured, but that such an arbitrary omnipotent Deity, might designedly make them such, as should be deceived in all their clearest perceptions.

[3] Ibid. 647. Wherefore we do not affirm, God to be so omnipotent or infinitely powerful, as that he is able to destroy or change the intelligible natures of things at pleasure; this being all one, as to say, that God is so omnipotent and infinitely powerful that he is able to destroy or to baffle and befool his own wisdom and understanding; which is the very rule and measure of his power.

[4] Fr. W. 34. But as it is very absurd to make active indifference blindly and fortuitously determining itself, that is active irrationality and nonsense, to be the hegemonic and ruling principle in every man.

... a will that is nothing else but will, meer impetus force and activity without any thing of light or understanding, a will which acts both it knows not why or wherefore, and even it knows not what...

[5] 1. Sermon 27. Another mistake, to our likeness is, ... when we make him nothing but a blind, dark, impetuous selfwill, running through the world; such as we ourselves are furiously acted with, that have not the ballast of absolute goodnesse to poize and settle us.

God is bound to restrict His omnipotence to meaning and purpose, means that He cannot deny His own divinity.

Above the *νοῦς*, as its first *νοητόν*, stands the *ἓν ἀγαθόν*, in abounding goodness communicating itself to the *νοῦς* [1]. The *ἓν ἀγαθόν* is the final consummation of the perfection of God, for the glory of God is His supreme goodness [2]. Without goodness, no majesty [3]. To believe in God means to believe and to hope all goodness, all beauty, all loveliness which a man can ever imagine; and then to know that this is only the smallest particle of the true beauty and goodness which is His [4].

[1] Fr. W. 50. There is a nature of goodness, and a nature of wisdom antecedent to the will of God, which is the rule and measure of it.

[2] T. I. S. 205. ... What the nature of God is, that he is properly, neither power nor knowledge (though having the perfection of both in him) but love.

Ibid. 664. Atheistick infidelity must needs on the contrary be, a certain heavy diffidence, despondence and misgiving of mind, or a timorous distrust and disbelief of good, to be hoped for, beyond the reach of sense.

[3] Ibid. 202. For knowledge and power alone, will not make a God. For God is generally conceived by all to be a most venerable and most desirable being. Whereas an omniscient and omnipotent arbitrary Deity, that hath nothing either of benignity or morality in its nature to measure and regulate its will, as it could not be truly august and venerable according to that maxim, "SINE BONITATE NULLA MAJESTAS"; so neither could it be desirable, it being that which could only be feared and dreaded, but not have any firm faith or confidence placed in it.

T. I. S. 203. (Knowledge is higher than power; and higher than knowledge and power is goodness.)

[4] Ibid. 661. Faith, hope and love, do all suppose an essential goodness in the Deity. God is such a being, who if he were not, were of all things whatsoever most to be wished for. It being indeed no way desirable ... for a man to live in a world, void of a God and Providence. He that believes a God, believes all that good and perfection in the universe, which his heart can possibly wish or desire. It is the interest of none, that there should be no God.

Ibid. 889. God is such a being, as, if he could be supposed not to be, there is nothing which any who are not desperately engaged in wickedness, no not atheists themselves, could possibly more wish for, or desire. To believe a God, is to believe the existence of all possible good and perfection in the universe; it is to believe, that things are as they should be, and that the world is so well framed and governed, as that the whole system thereof, could not possibly have been better.

There is nothing, which cannot be hoped for, by a good man, from the Deity; whatsoever happiness his being is capable of; and such things as eye hath not seen, nor ear heard, nor can now enter into the heart of man to conceive. Infinite hopes lie before us, from the existence of a being infinitely good and powerful, and our own

Those who restrict the goodness of God for the sake of irrational omnipotence, deny His majesty also; and if they preach an arbitrary, almighty God, they do more harm, than if they denied His existence altogether, and put in His place dead matter, as the source of life and understanding[1].

God is God, because He is goodness[2]. If He could act against His own goodness and truth, He would not be God.

Thus we find within the Trinity a subordination in the sense, that Power has its perfection in Wisdom, and Wisdom its consummation in the supreme Goodness of God. But since Goodness, Wisdom, and Power are divine, and as God, one perfect unity, Cudworth can say with Descartes that the will of God IS His Being, wisdom, and his supreme goodness[3]. Thereby, subordination again loses absolute significance and is taken up into the ultimate oneness and inexistence of the Persons of the Trinity. The goodness of God is the will to fair justice, and justice is wisdom. The decisive difference from Descartes is, that this unity is realised in Trinity, and essentially IS Trinity.

Not only knowledge, but also morality is, in a twofold way, founded upon the Trinity.

souls immortality: and nothing can hinder or obstruct these hopes, but our own wickedness of life. To believe a God, and do well, are two, the most hopeful, cheerful and comfortable things, that possibly can be.

[1] Ibid. 203. And indeed an omnipotent arbitrary Deity, may seem to be in some sense, a worse and more undesirable thing, than the Manichean evil God; for as much as the latter could be but finitely evil, whereas the former might be so infinitely.

And it also seems less dishonourable to God, to impute defect of power than of goodness and justice to him.

[2] 1. Sermon 26. That God is therefore God, because he is the highest and most perfect good: and good is not therefore good, because God, out of an arbitrary will of his, would have it so. Whatsoever God doth in the world, he does it as it is suitable to the highest goodnesse; the first idea, and fairest copy of which is his own essence.

[3] Ibid. 37. God's power displaied in the world, is nothing but his goodnesse strongly reaching all things, from heighth to depth, from the highest heaven, to the lowest hell: and irresistibly imparting it self to every thing, according to those severall degrees in which it is capable of it. (MODO RECIPIENTIS)

For the root of all power, is goodnesse.

T. I. S. 406. For his will is his essential goodness, so that his will doth not follow his nature but concurr with it.

a) Since goodness is the highest attribute of God, we attain a higher participation in God by holiness than by knowledge.

b) In a similar way to that in which the divine will depends on wisdom, our actions also are guided by the knowledge of the immutable Natures of Good and Evil. This knowledge can be attained by different ways, yet, to become explicitly conscious, it must always pass through the mediation of discursive reason.

The foundation of knowledge and morality in the Trinity gives to Cudworth's philosophy a tremendous inner life and power. It also forms an effective defence against dogmatism[1], because the strict transcendental foundation of knowledge puts an end to any form of autonomy of reason, and discloses the claim of absolute certainty in any finite knowledge as an illusion.

Cudworth conceives the relationship of the Persons of the Trinity as a pure act, whose object is perfect and immutable. And, by virtue of the integral oneness of the three Persons, act and content are in the Trinity perfectly one. This same oneness of act and content he conceives also within each separate hypostasis, as we have seen it in the *νοῦς*. Act and content, subject and substance, energy and rest, are thus inseparably one in a double sense within the Trinity. By this conception, Cudworth finally overcomes the statics of the Cartesian dualism in a more convincing way than by the hypothesis of Plastic Nature. But he also left far behind the Neo-Platonic conception of the *νοῦς*, as a static intelligible relation system of the universe, which he adopted and transformed much in the same way, as he had done the dualism of substances. In this he was far in advance of his time.

We realise this, when we again cast a passing glance at the system of *Hegel*. For Hegel also the Trinity is the pulse of knowledge, since the dialectical, speculative progression from thesis through antithesis to synthesis, is transcendentally founded upon the Trinity, which Hegel interpreted in the following way. The *ἕν* sends itself out into the multiplicity of the *λόγος* as its antithesis, and returns to the synthesis, the unity in multiplicity, in the Holy Ghost. Each finite act of thinking, which progresses in the same way from thesis through antithesis to synthesis, is achieved within the Trinity, according to Hegel. It is

[1] Ibid. 199. For there may be a latitude allowed in theism.

lifted up into the life of the Trinity. Or, from the point of view of the Trinity, it could be said that the life of the Trinity realises itself through the finite speculative thinking, because, according to Hegel, it is the Absolute Spirit, who, through finite minds, in infinite dialectical speculative progression, gradually gains more perfect consciousness of its own content.

The decisive difference here again is found in that Hegel believes that he could overcome the gulf between finite and infinite, between perfect and imperfect, by finite dialectical speculative thought, which, however, he conceived as itself a constitutive part of the pure act of thinking of the *νοῦς*, and as ultimately one with it.

Cudworth, following Plato, holds to the perfect transcendence of the Trinity, not however, as opposed to the finite, for this would necessarily make it relative. Finite reason and will are transcendentally founded on the life of the Trinity, in the sense that the Trinity is their "terminus a quo" and their "terminus ad quem". All finite life is thus enclosed and taken up into the life of the Trinity, yet the gulf between finite and infinite, is in no way illegitimately passed from the side of the finite.

On the other hand, the participation of finite life in the life of the Trinity implies the bridging of the gulf by the Perfect. A participation of the imperfect in the Perfect is possible only if the Perfect condescends to the imperfect; and thus the self-limitation of God is seen to be the very condition of every finite act of thought and will. It was already the condition of the creation of the world, since, as we have seen, divine creative power, can act "ad extra" only with time, because finite Being cannot exist in the mode of the eternal, but would necessarily break under the full pressure of an almighty creative will.

The significance of the Trinity, and of the condescension of God, revealed in the Incarnation, in the philosophy of Cudworth, will appear in a new and still clearer light, when we pass to his exposition of ethics, and therewith enter the innermost circle of his thought.

Chapter V

ETHICS

The significance of Ethics in the philosophy of Descartes

Cudworth has shown that Descartes, by rejecting the final cause and the Trinity, destroyed the foundations of knowledge altogether. It is therefore not surprising that in Ethics, the centre of his own thought, Cudworth no longer finds in Descartes an adequate opponent.

Descartes' chief interest was free scientific research on sure metaphysical foundations, from which it was possible for knowledge to progress with absolute certainty. Yet moral purpose is not altogether absent from his philosophy; it is clearly felt, although it is not given sufficient philosophical justification within his system.

Ethics, in Descartes' philosophy, appears in three different aspects.

1. Descartes establishes preliminary moral maxims and claims that, for practical reasons, they must be acknowledged as binding, even before their evidence can be logically deduced. They serve primarily as a safeguard against methodical, theoretical doubt, lest practical life be undermined by its critical power. Further, they are to procure peace of mind and leisure (σχολή), which are indispensable for good scientific work. Descartes stresses the fact that, without true peace of mind, we cannot use our cognoscitive powers rightly. Here, therefore, we may discern that Descartes himself recognises a certain dependence of knowledge upon moral disposition.

This claim for preliminary morality, as a safeguard against doubt, indirectly supports Cudworth's argument that the methodical doubt is no true doubt at all. True doubt always originates from experience, and never from necessary reason, and there is no protection from it, except to overcome it by thought.

The same objection can, in a certain sense, be brought against Hegel also, when he asserts that Reason itself, "behind the back of consciousness", creates the antithesis in order to be able to solve it in the synthesis.

Cudworth, moreover, disputes the point, that maxims, which have sprung from arbitrary, utilitarian considerations, can ever keep their binding force under the pressure of practical life, since they obviously have no categorical obligation of themselves; "they would not even have cobweb strength". The philosopher cannot find true peace of mind, either in accidental outward circumstances, or in any specific psychological dispositions, but solely, when he rises up to the objectivity of the *νοῦς*, and, in the act of pure thinking, effaces and transcends, in true self-denial, his own finite individuality.

2. Descartes planned to draw up a whole system of sciences which he proposed to call the Tree of Knowledge, and in which he attributed to Ethics, as the highest knowledge, the supreme place. He believed that it could be deduced "clare et distincte" from the metaphysical foundations of knowledge, and that from this logical deduction the categorical validity of the moral maxims will result. But unfortunately, as Cudworth says, "By an undiscerned tang of mechanical atheism hanging about him" [1], Descartes was rather drawn to scientific research, and therefore never achieved his "tree of knowledge".

One might ask, on what foundation an absolutely binding morality could be built up in a universe which has no "terminus ad quem", and with a science, in which time is reversible and has no direction and purpose. Or is the denial of the final cause meant only as a scientific, though illusionary, method of research, which can, at any time, be again abandoned?

A further difficulty for the foundation of morality arises from the arbitrary character of the Natures of Good and Evil, which according to Descartes were created by God, and cannot therefore have in themselves the force of obligation. This problem is not solved by referring to the eternity and the immutability of the decrees of God, because "immutable arbitrariness" is a contradiction in terms, and therefore nonentical. In its essence, arbitrariness is accidental and finite, and cannot therefore be attributed to an eternal will.

Cudworth fought to the end the argument against arbitrary Truth.

[1] Ibid. 147.

In the battle for the eternal immutable Nature of the Good he regards the champions of predestination to evil as more harmful opponents.

3. Descartes reveals his truest thought concerning morality in his letters to the Princess Elizabeth. There, in simple and beautiful words, he says that it is the supreme and noble task of philosophy to show to men the way to true peace and happiness in the midst of the adversities of practical life. We shall attain it, he says, if in regular meditation, we train ourselves to see and understand that our true self is incorporeal, and cannot be harmed by any adversity or death. Slowly and patiently we can learn to distinguish true and lasting values from those which are passing. We can learn to rise from deceptive sensations to the clarity of pure thought. Then we shall be able to overcome and leave behind our emotions and passions, which, even more than the outside world, are apt to disturb our peace.

We see here that morality is for Descartes a true philosophical aim, although he never tried to give it a logical foundation, as he had originally planned.

For Cudworth, the finite soul is not a mode of the substance of Cogitation, it has therefore no unconditioned immortality in itself, but it is exposed to every chance in the flux of Becoming. Therefore the contemplation of its true Nature could never lift up our soul beyond Becoming and transitoriness. The assurance of our immortality is granted us only in the degree, in which we transcend ourselves, in an act of will, for the sake of participating in thought and will in the divine mind and the divine will.

All these different aspects of morality in Descartes' philosophy confirm Cudworth's opinion that Descartes in truth endeavoured to put the metaphysical principles in human reason, although he still acknowledged a certain dependence upon a higher principle. Yet the main current of his thought points clearly to the autonomy of the finite substance of Cogitation.

The examination of Descartes' conception of morality does not show us any new aspects, neither does it lead to the central problems. A sound philosophical foundation of Ethics is impossible, unless the Absolute is conceived to embrace all finite Being as its beginning and end, its first cause and purpose. Ethics is conceivable only in strict dependence on an absolute principle, and together with a conception of knowledge founded on transcendental principles. All the lines of Logic and Theology converge in Ethics, and here is the key, without which Cudworth's philosophy can never be seen in its true proportion and beauty.

The metaphysical foundations of Ethics

In Cudworth's philosophy Ethics is founded on two different planes, objectively in the Trinity, and subjectively in Freewill.

1. The transcendental foundation of Ethics. Before we set out to examine the transcendental foundation of Ethics, we do well to call to mind the initial ambiguity that Goodness, Wisdom, and Power of God are conceived as attributes of God, and are at the same time seen to represent the persons of the Trinity. As attributes of God they are considered as inadequate expressions of the one conception of the Perfect; yet at the same time they represent the essence of the Persons of the Trinity so perfectly that they can be identified with them.

Cudworth sees Ethics founded on the Trinity in a twofold way.

a) By the fact that the highest hypostasis of the Trinity is the *ἓν ἀγαθόν*, the supreme Good. The *ἓν ἀγαθόν* reveals itself in the idea and, vice versa, the *νοῦς* IS towards the *ἓν ἀγαθόν*. Goodness and Wisdom of God can thus never be separated from each other[1]. There is therefore no true knowledge which is not ultimately knowledge of the supreme Good[2]. And in the knowledge of the ideas, the knowledge of the supreme Good is necessarily always implicit. There is, therefore, no knowledge of Being or fact which is not also knowledge of its meaning and its final cause. In other words, with the knowledge of what things are is inseparably connected the knowledge of " how it is good for the things to be ".

[1] Fr. W. 49–50. God is a nature of infinite love, goodness, or benignity, displaying itself according to infinite and perfect wisdom, and governing all creatures in righteousness, and this is liberty of the Deity.

[2] Cf. T. I. S. 874. God is *μέτρον πάντων*, an impartial ballance, lying even equal and indifferent to all things, and weighing out heaven and earth, and all the things therein, in the most just and exact proportions, and not a grain too much or too little of any thing.

Nor is the Deity therefore bound or obliged to do the best, in any way of servility ... but only by the perfection of its own nature, which it cannot possibly deviate from, no more than ungod it self.

The meaning and purpose of the things is their *προσῆκον*[1], which can also be symbolised by "proportion" and "symmetry" as the presupposition of the Being of all things. Meaning, purpose, and Being of the things are inseparably one. No Being can be conceived, unless it has meaning and purpose; unless it ultimately IS towards the supreme Good[2].

In this sense Ethics has priority over Logic, the knowledge of mere Being. On the other hand, Logic passes directly into Ethics, and both are inseparable from each other, because in all knowledge of ideas (and we can only think in ideas), Ethics is already presupposed and implicit, by virtue of the fundamental oneness of Goodness and Wisdom in God[3].

b) The other aspects of the transcendental foundations of Ethics spring from the "Natural imperium" of the *νοῦς* over the divine Will, and from the immutable eternal Natures of Good and Evil.

The divine Will is, in a twofold way, bound to the unchanging Goodness: by the "how" of its creative work it is bound to the supreme Goodness of God as the highest hypostasis; and in the "what" of its creative work it is bound, firstly, to the "Natures" of the things, which are their "perfection", secondly, to natural justice as idea in the divine *νοῦς*.

[1] 1. Sermon 50. The strong magick of nature, pulls and draws every thing, continually, to that place which is suitable to it, and to which it doth belong.

T. I. S. 881. Whereas were there but one kind of thing (the best), thus made; there could have been no musick nor harmony at all, in the world for want of variety. But we ought in the first place, to consider the whole, whether that be not the best, that could be made, having all that belongeth to it; and then the parts in reference to the whole, whether they be not in their several degrees and ranks, congruous and agreeable thereunto.

Plato. Tim. 87*c*.

[2] Fr. W. 29–30. But above all these (the particular goods), the soul of man hath in it a certain vaticination, presage, scent, and odor of one SUMMUM BONUM, one supreme highest good transcending all others, without which they will be all ineffectual as to complete happiness, and signify nothing, a certain philosophers' stone that can turn all into gold.

Now this love and desire of good, as good, in general, and of happiness, traversing the soul continually, and actuating and provoking it continually, is not a mere passion or HORME, but a settled resolved principle, and the very source, and fountain, and centre of life.

[3] Plato, Tim. 37*a*–*b*.

We have here the illustration for the double symbolism of the relationship within the Trinity by straight line and circle. More than that, it is a wonderful example of the flexibility of Cudworth's thought, due to the fact that he considered our means of expression as no more than symbols. We can never possess Truth as a whole, nor can we have it as absolute certainty once for all; we can only contemplate it from various particular aspects, whereby every new aspect, even if it should not precisely fit with what has been seen before, is welcomed, since it makes richer and more complete our inadequate and partial knowledge. The more aspects we can see, the nearer we get to the comprehension of the whole, and also the more truly we know that the Whole will always be beyond the grasp of our finite capacity.

Cudworth conceives will as a restless desire for the supreme Good[1]. And from this point of view the *ἓν ἀγαθόν* can be seen as the final cause even within the Trinity.

All knowledge is founded on the fact that the will, in its creative action, is bound to the eternal Natures, and thereby also to the Good, the *προσῆκον*. It can therefore only bring forth Goodness, proportion, symmetry; it can create only beauty. Goodness and beauty are thus revealed as the fundamental presupposition for all Being and knowledge in every respect[2]. The creative will is also bound to the idea of Natural Justice, and thus dependent on the *ἀγαθόν*, both, directly and

[1] Fr. W. 28. Wherefore, we conclude that the spring and principle of all deliberative action, can be no other than a constant, restless, uninterrupted desire, or love of good as such, and happiness. This is an ever bubbling fountain in the centre of the soul ... a spring of motion, both a PRIMUM AND PERPETUUM MOBILE in us ... an everlasting and incessant mover.

Ibid. 28–9. God, an absolutely perfect being is not this love of indigent desire, but a love of overflowing fulness and redundancy, communicating itself.

But imperfect beings, as human souls, especially lapsed, by reason of the PENIA which is in them, are in continual inquest, restless desire, and search, always pursuing a scent of good before them and hunting after it.

Muirhead, Manuscr. 66. Cudworth, like a true Platonist, finds this "natural and necessary inclination" (of the will) in the idea of "good". We do not will merely because we will, but it is always for the sake of some good... There is talk of hating good as good and loving evil as such. But it is rather to be thought impossible that there should be any such being in the world.

[2] Plato, Phileb. 64*d*–65*a*.

through the mediation of the divine νοῦς[1]. In a threefold respect the Will, in its creative work, is directed towards the supreme Good, and herein is its perfection and power. Since perfect Goodness is perfect Reality, the higher the degree of reality is of that which the Will is able to create, the greater also is its power.

Holiness is thus revealed as the principle of the divine life, and also of all finite life in general. Righteousness is the rule of the divine Will and of all His creative work.

Here begins Cudworth's wonderful and memorable refutation of the Calvinist doctrine of divine omnipotence. The philosophical argument runs on parallel lines to the refutation of the Cartesian theory of created Truth. Cudworth accuses the Calvinists for proclaiming the glory of God as an arbitrary, absolute will, and for putting arbitrary decrees of moral values in the place of the eternal, immutable Natures of Good and Evil, in analogy with Descartes' assumption of created Truth. Descartes thereby destroyed the principles of knowledge, but, because knowledge is inseparably united to Good, he indirectly destroyed the metaphysical foundation of morality. Calvin, on the other hand, destroyed knowledge by breaking up the metaphysical foundations of morality, since knowlege, fundamentally, is always knowledge of the Good. The result is the same; whenever omnipotence is understood as an arbitrary will, it is contingence and chance which determine the Goodness and Wisdom of God[2], and make it finite, since nothing, which is exposed to chance, can bear in itself eternity and immutability.

[1] Fr. W. 52. All will is generally acknowledged to have this naturally and necessarily belonging to it, to be determinated in good, as its object; it being impossible that any intelligent being should will evil as such.

Ibid. 54. Though all things in the universe had not been arbitrarily made such as they are, but according to the best art and wisdom, yet were they not therefore less διὰ τὴν θέλησιν θεοῦ, for the will of God. It being his will to make them according to his wisdom; or to order all things in number, measure, and weight (Wisdom 11,20).

[2] T. I. S. 872–3. However some modern theists ... concluding the perfection of the Deity, not at all to consist in goodness; but in power and arbitrary will only.

But the prerogative of a being irresistibly powerful, to have a will absolutely indifferent to all things, and undetermined by any thing but it self; or to will nothing because it is good, but to make its own arbitrary or contingent and fortuitous determination, the sole reason of all its actions, nay the very rule or measure, of goodness, justice, and wisdom it self.

Cudworth describes in most vivid terms the conception of God which underlies the doctrine of predestination to evil, if it is carried to its conclusion. It is indeed rather the conception of an almighty terror than of God, a terror which in crazy omnipotence, without perfection, madly dashes through the world[1]. Holiness then is incongruous and alien to God, for He is not God, if there were any evil which He could not achieve[2]. Neither is there any evil which, by the decree of God, could not be declared as good. God thus is made the only agent for evil as well as for good, truly a kind of World-Soul.

What else can predestination mean than that God, in final moral indifference, chooses a few who, unhealed from within, are perforce declared sharers in the divine beatitude by an outward decree, as if the soul, apart from holiness, could experience any joy of perfection[3]?

[1] T. I. S. 873. But what do these theists here else, than whilst they deny, the fortuitous motion of senseless matter, to be the first original of all things, themselves in the mean time enthrone fortuitousness and contingency, in the will of an omnipotent being, and there give it an absolute soveraignty and dominion over all?

Certainly, we mortals could have little better ground, for our faith and hope, in such an omnipotent arbitrary will as this, than we could have in the motions of senseless atoms, furiously agitated; or of a rapid whirlwind. Nay one would think, that of the two, it should be more desirable, to be under the empire of senseless atoms, fortuitously moved, than of a will altogether undetermined by goodness, justice, and wisdom, armed with omnipotence; because the former could harbour no hurtful or mischievous designs, against any, as the latter might.

[2] Mor. 11. Now the necessary and unavoidable consequences of this opinion are such as these ... that holiness is not a conformity with the nature of God.

Ibid. 9–10. Whence it follows unavoidably, that nothing can be imagined so grossly wicked, or so fouly unjust or dishonest, but if it were supposed to be commended by this omnipotent Deity, must needs upon that hypothesis forthwith become holy, just and righteous.

That there is no act evil but as it is prohibited by God, and which cannot be made good if it be not commanded by God.

Therefore that God could not be God if there should be any thing evil in its own nature which he could not do.

[3] 1. Sermon 27. God, who is absolute goodnesse, cannot love any of his creatures, and take pleasure in them, without bestowing a communication of his goodnesse and likenesse upon them.

2. Sermon 208 ... It being a thing utterly impossible, that those undefiled rewards of the heavenly kingdome should be received and enjoyed by men in their unregenerate and unrenewed nature.

To worship such a God, framed after the image of man's own wicked heart, is worse indeed, than to make idols of wood or stone, which at least cannot do much harm[1]. And it would be less harmful with the atheists to declare dead matter to be the first principle, instead of an almighty irrational arbitrariness which, in blind fury, destroys that which it has created. By the power of his great vision, Cudworth discloses this travesty and betrayal, not only of the Christian Faith, but of all religions, and he adds that they who have drawn out these monstrous consequences are no more to blame than they who laid the premises; at least they were honest enough to show all their cards[2].

Cudworth exposes the barbarous illusion at the bottom of this doctrine, by proving that arbitrariness destroys the very essence of power. Perfection of power is found in its being guided by perfect Wisdom and perfect Goodness alone. "Power for evil" will always in the end disclose itself as impotence[3]. For by doing evil it automatically turns against itself.

In a most impressive and wonderful way, Cudworth shows in his theory of the state, what is the practical significance of morality, thus founded in the divine *νοῦς*. We find his theory of the state developed in the refutation of the Leviathan of Thomas Hobbes. J.H. Muirhead calls it the fairest heritage of British freedom and loyalty towards the state[4].

The theory of the state founded on Natural Justice[5]

"God and Nature create the state[6]", without them men labour in vain. "Reason establishes every throne[7]." The will can act only to-

[1] 1. Sermon 23–4.

[2] (We are reminded of Hume's "crisis", which could almost be understood as a conversion from Calvinism to atheism.)

[3] T.I.S. 873. But this irrational will, altogether undetermined by goodness, justice and wisdom, is so far from being the highest, liberty, soveraignty and dominion; the greatest perfection, and the divinest thing of all; that it is indeed nothing else but weakness and impotency it self, or brutish folly and madness.

[4] Muirhead. The Platonic Tradition in Anglo-Saxon Philosophy 28.

[5] T.I.S. 893 f.

[6] Ibid. 896. Had not God and nature made a city; were there not a natural conciliation of all rational creatures, and subjection of them to the Deity, as their head,

wards the Good; authority and obligation are founded on the Absolute, on the supreme Goodness of God, on Natural Justice; and in that the will of God is guided by His Goodness and Wisdom[1]. The anticipations of authority and obligation must be presupposed in rational beings as such, for without them, no command could be given with binding obligation. The first obligation could never be commanded by any positive law, because, without the anticipation of it, it could not be recognised as such[2]. Obligation can only BE, and ultimately it is always obligation to the supreme Good.

The state is rational, both by the fact of its foundation, and also by the way, in which it is realised practically.

1. States can be formed, because men, as rational beings are by nature social beings. Their social character springs from their being rooted in the one divine mind, participating in the supreme Order and Purpose of the Whole, which itself IS the manifestation of the supreme Good[3].

had not God made ... superiority and subjection, with their respective duty and obligation, men could neither by art, or political enchantment, nor yet by force, have made any firm cities or polities.

[7] 1. Sermon Pref. For justice and righteousnesse are the establishment of every throne, of all civil power, and authority; and if these should once forsake it, though there be lions to support it, it could not stand long.

[1] Cf. T. I. S. 895. Indeed the ligaments, by which these politicians would tie the members of their huge Leviathan, or artificial man together, are not so good as cobwebs; they being really nothing, but meer will and words. For if authority and sovereignty be made only by will and words, then is it plain, that by will and words, they may be unmade again at pleasure.

[2] T. I. S. 698. It remaineth therefore, that conscience and religious obligation to duty is the only basis, and essential foundation of a polity or commonwealth; without which there could be no right or authority of commanding in any sovereign, nor validity in any laws.

Ibid. 697. Pacts and covenants without this obligation in conscience, are nothing but meer words and breath. The laws and commands of civil sovereigns, do not make obligation, but presuppose it, as a thing in order of nature before them, and without which they would be invalid.

[3] T. I. S. 896. Which bound or vinculum can be no other, than natural justice; and something of a common and publick, of a cementing and conglutinating nature, in all rational beings; the original of both which, is from the Deity. The right and authority of God himself is founded in justice; and of this is the civil sovereignty also a certain participation.

The state is based upon the community of rational beings, which has its transcendental foundation in the *νοῦς*. In this sense the state is "one whole", a unity in multiplicity. If the particular units had by nature a tendency to disunity, and were not by nature related to and connected with each other, no artificial contract could unite them into one organic whole. Such a state could be kept in existence only by senseless violence which is not founded on Truth. A most precarious state of affairs would result, since arbitrary political power inevitably sanctions, by its own example, rebellion, and is fundamentally hostile to the formation of the state. Cudworth says that the "force of lions" could not prevent the citizens of an artificial state from lapsing into their natural isolation. But world history proves, on the contrary, a natural tendency to form communities and states[1]. Even an anarchist ideal of state is thought of as a form of common life.

As "one whole" the state is necessarily founded upon reason, for reason alone has the power of gathering multiplicity into the unity of the final cause. The state can be realised only because its members by nature live a community life. This "natural community life" necessarily precedes every common order and purpose in any form of common life. Common life could never be created artificially; it also can only BE, as Truth IS and original life IS.

Since the state rests upon this natural and reasonable community life which, a priori, IS, therefore no outward violence is needed in its constitution, preservation, and government. It is natural to man to unite under a common purpose, and it intrinsically accords with man's own will, which, by its nature strives after that which is reasonable. A state can therefore be governed from within, "gracefully" one might almost say.

Cudworth mentions one exception, where the application of force is advisable, that is for the counteracting of irrational tendencies, when

[1] Ibid. 895. Lastly, were civil sovereigns and bodies politick, meer violent and contra-natural things, then would they all quickly vanish into nothing, because nature will prevail against force and violence: Whereas men constantly every where fall into political order, and the corruption of one form of government, is but the generation of another.

they resist the natural imperium of the sanctioned government, and rebel[1].

2. Cudworth distinguishes laws with categorical obligation from laws with only hypothetical obligation[2]. The values of categorical obligation are, in analogy with pure ideas, the foundation of the values of hypothetical obligation. Categorical values carry universal obligation to absolute obedience for rational beings as such, in all places and at all times. Cudworth does not give an account of the absolute values, yet it is easy to divine from the context that they are all modifications of the love, which in the life of the Trinity is perfectly realised. Since love spontaneously unites multiplicity into unity, a loving life is the only true and reasonable life. The relationship between the Persons of the Trinity is the highest realisation of reason, consequently the highest reality. It is perfect love.

Every positive law is accidental, and has but indirect obligation. All the laws of state belong to this category. They have force of obligation in so far as they are related to the pure values of Natural Justice. A command, which in itself is indifferent in value, can for instance acquire categorical obligation by the fact that it is promised. This, however, does not change its character of indifference as regards the content. Laws of state can, in like manner, acquire categorical obligation by the fact that they are part of the sphere of civil duty and obedience. New values are not created by any positive law. Any command carries the force of law only in so far, as it is founded upon Natural Justice

[1] Ibid. 896.

[2] Mor. 21. Wherefore the difference of these things lies wholly in this, that there are some things which the intellectual nature obligeth to of it self, and directly, absolutely and perpetually, and these things are called naturally good and evil; other things there are which the same intellectual nature obligeth to by accident only, and hypothetically, upon condition of some voluntary action either of our own or some other persons, by means whereof those things which were in their own nature indifferent, falling under something that is absolutely good or evil, and thereby acquiring a new relation to the intellectual nature, do for the time become such things as ought to be done or omitted, being made such not by will but by nature.

Ibid. 20. ... even in positive laws and commands it is not meer will that obligeth, but the natures of good and evil, just and unjust, really existing in the world.

itself[1], which is the absolute, immutable and eternal norm of action, and the only standard of judgment. It is therefore impossible to give an unjust command with force of obligation, because obligation is founded upon Justice itself[2].

Incited by the outward positive law, we can awaken in ourselves the pure values by virtue of the anticipation of Natural Justice. The cognoscitive power for values is conscience[3]. A command can reach us at all, only because, incited by the experience of outward law, we self-actively realise the pure values in ourselves, both as regards the content and the readiness for their realisation. In our conscience we participate therefore in a twofold way in the life of the Trinity. As regards the content, we participate in the Natural Justice and Pure Values; as regards the will, we participate in the perfect realisation of justice within the Trinity.

[1] Ibid. 25–6. Wherefore in positive commands, the will of the commander doth not create any new moral entity, but only diversly modifies and determines that general duty or obligation of natural justice...

... It is not possible that any command of God or man should oblige otherwise than by virtue of that which is naturally just.

For meer will cannot change the moral nature of actions, nor the nature of intellectual beings.

[2] T. I. S. 897. Whereas the goodness of justice or righteousness is intrinsecal to the thing it self, and this is that which obligeth, (and not any thing forreign to it) it being a different species of good from that of appetite and private utility, which every man may dispense withal.

Now there can be no more infinite justice, than there can be infinite rule, or infinite measure. Justice is essentially a determinate thing.

If there be any thing in its own nature just, and obliging ... then must there of necessity be something unjust or unlawful, which therefore cannot be obligingly commanded by any authority whatsoever.

[3] Cf. Muirhead, Manuscr. 65.

Fr. W. 31. Above all these (the lower powers of the soul, fancy, imagination, sudden passions and hormae and commotions called concupiscible and irascible) is the dictate of honesty, ... of conscience, which often majestically controls them, and clashes with the former; this is necessary nature too, being here the hegemonic, sometimes joining its assistance to the better one, and sometimes taking part with the worser against it.

This conscience is universal, both in act and content. It is the only power capable of leading us beyond mere private interests into true citizenship[1].

In analogy with sense-perception, of which we could never become conscious at all without the self-activity of reason, outward law could never reach us, if we had no inner corresponding faculty within us. Only spirit can read spirit, and only spirit can, through the mediation of sense experience, hear and obey spirit. Violence is not needed for the execution of just laws; laws oblige from within[2].

The categorical obligation originates in the fact that conscience is founded in the objectivity of the Trinity. Therein, every moral act transcends its accidental character and is taken up into the sphere of the absolute[3]. Moral error assumes the proportions of sin in an ab-

[1] T. I. S. 898. (This conscience is universal according to the universal rule of natural justice. In act it is universal in so far as it can be actuated only by participation in the divine will. But in so far as this actualisation can be achieved only in particular finite beings, it is finite and contingent according to act *and* content, and can err.)

Cf. ibid. 898. Wherefore conscience also, is in it self not of a private and partial but of a publick and common nature; it respecting divine laws, impartial justice, and equity, and the good of the whole, when clashing with our own selfish good; and private utility. This is the only thing, that can naturally consociate mankind together, lay a foundation for bodies politick, and take away that private will and judgment according to men's appetite and utility, which is inconsistent with the same.

Ibid. 898 f. It is true indeed, that particular persons must make a judgment in conscience for themselves (a public conscience, being nonsense and ridiculous) and that they may also erre therein; yet is not the rule neither, by which conscience judgeth, private, nor it self unaccountable...

Ibid. 897. And could civil sovereigns utterly demolish and destroy, conscience and religion in the minds of men, (which yet is an absolute impossibility) they thinking thereby to make elbow-room for themselves, they would certainly bury themselves also, in the ruins of them.

[2] Cf. ibid. 895. Wherefore since it is plain, that sovereignty and bodies politick can neither be meerly artificial, nor yet violent things, there must of necessity be some natural bond or vinculum to hold them together, such as may both really oblige subjects to obey the lawful commands of sovereigns, and sovereigns in commanding to seek the good and welfare of their subjects.

[3] T. I. S. 897. The right and authority of God himself, who is the supreme sovereign of the universe, is also in like manner bounded and circumscribed by justice.

solute Court of Justice. With this argument Cudworth answers the objection that eternal punishment is disproportionate to accidental sin. There is no such thing as contingent sin. Positive laws, if they were not ultimately founded on the absolute, could call forth only accidentally good actions, done for reward and not for the sake of pure Justice.

In the imperial power, Cudworth sees a participation of a higher order in God as the supreme Ruler of the world. The king alone is therefore given power over life and death, and absolute obedience to him is a primary command of Justice[1]. Since Justice is founded in the divine *νοῦς* itself, there can be no rational being which is indifferent to the Good[2].

The relations between Logic and Ethics

Ethics is, as we have seen, founded in the Absolute in a threefold way.

a) In the *ἓν ἀγαθόν*, the supreme Good, the highest hypostasis.

b) In the relationship between the Persons of the Trinity, which can be symbolised by inexistence as a perfect and immediate realisation of the supreme Good.

c) In the idea of Natural Justice, the immutable Order of Pure Values.

We find here precise analogies with the metaphysical foundations of knowledge. Thus, the pure values are at the same time pure ideas; the anticipations of the pure values in the human mind, and their self-active awakening in conscience, when incited by the experience of positive law, are parallel to the awakening of pure ideas in the intellect. In conscience we transcend our own individuality and have part in the universal. Thereby we attain a higher faculty of communication, not only in thought, but also in will. We are rendered capable of corporate actions for common purposes. As all men at all times can awaken the same pure ideas, so they can awaken the same immutable values in themselves.

God's will is ruled by his justice, and not his justice ruled by his will; and therefore God himself cannot command, what is in its own nature unjust.

[1] Ibid. 896.

[2] Mor. 26–7.

Cudworth makes use of this parallelism of thought and will, when he assumes that thought and life are ultimately the same. The principle of life is the will; thought and will are inseparably bound together, not only through their common foundation in the transcendent, but also in their common source in the finite subject. This leads to the problem of freewill, which Cudworth examines with uncommon candour and insight.

We discover still more intimate coherence between Logic and Ethics, if once more we consider all we have been saying of the ideas as Being, meaning, and purpose of things. The Meaning and purpose of the things is manifest in their proportions within themselves, and in the symmetrical Order of the Whole, of which they are part, together with all other things in the universe. Proportion is the beauty of things, and it is also their Being; in so far as they can preserve their symmetry, they ARE[1]. Their reality is their beauty. The One divine Order of the Whole is the manifestation of the supreme Good. One can therefore say that the oneness of Love is also the oneness of Truth; the ideas are therefore also the goodness of things, as well as their Truth and Being, and in so far as things are able to be images of the ideas, they are such as it is good for them to be. The knowledge of ideas implies at once knowledge of Being, of Truth, of beauty, and goodness. It is for this reason that Ethics cannot be severed from Logic[2].

This is of great importance for the finite act of knowledge. It means that a man can attain true knowledge only, if he is directed with his whole heart towards the supreme Good[3]. In other words, philosophy makes demands on the whole personality. From another side, it could be said that every newly gained knowledge also strengthens the will towards the Good.

Knowledge of Truth, beauty, and goodness is ultimately knowledge of the One Perfect, the Whole, apprehended in different harmonies; the One, to which by thinking we can never attain in its perfection, yet

[1] Plato, Res Publ. 505*a–b*.
[2] T.I.S. Pref.
Plato, Phaedo 97*c–d*.
Nom. 891*e*.
[3] Plato, Phaedr. 250*a*.

its *μαντεία* lives in us[1]. In the transcendent One all the lines of thought, will, and life finally converge.

Truth and beauty are the *ἓν ἀγαθόν* in its immediate self-revelation. Ethics has therefore a natural imperium over Logic and Aesthetics. It includes both. Our supreme participation in the *ἓν ἀγαθόν* does not lie in knowledge of Truth, nor in contemplation of beauty, but in holiness, which is the realisation of Love[2].

The foundation of Ethics in the finite subject

Freewill

The problem of freewill attracted Cudworth in a very exceptional way. During his later years especially he wrestled almost exclusively with this problem. But alas, the greatest part of his work still lies buried in his manuscripts in the British Museum. His great work, of which we have only "the fragment", the first volume, he intended to consist of three volumes and to describe the intelligible universe from the various points of view obtained from the study and refutation of every form of Determinism in science, philosophy, and religion; or, as Cudworth puts it, of "the mechanical 'atomical' Fate"; "the pantheistic Stoical Fate"; "the theological Calvinistic Fate". According to Cudworth, no morality or religion whatever would be possible, if man were not the object of divine justice, capable of receiving reward and punishment. For this, a certain sphere of free choice must be granted to him.

Perfect Will is the source and standard of finite will. This latter can

[1] T. I. S. 646. And though we, being finite, have no full comprehension and adequate understanding of this infinity and eternity (as not of the Deity) yet can we not be without some notion, conception and apprehension thereof, so long as we can thus demonstrate concerning it, that it belongs to something, and yet to nothing neither but a perfect immutable nature.

[2] 1. Sermon 68–9. Holinesse, whereever it is, though never so small, if it be but hearty and sincere, it can no more be cut off and discontinued from God; than a sunbeam here upon earth can be broken off from its entercourse with the sun and be left alone amidst the mire and dirt of this world.

Holinesse is something of God, whereever it is; it is an efflux from him, that always hangs upon him, and lives in him.

God cannot draw a curtain between himself and holinesse.

act only in so far as it participates in the Perfect Will[1]. Perfect Will is a Will guided by Perfect Wisdom; it is an intrinsically rational Will; to say that it is directed towards the ideas means that it is directed towards Being, Truth, Beauty, and Supreme Goodness. Its perfection and freedom lie in its perfect orientation towards Wisdom, and its ultimate oneness with it. Perfect freedom is the freedom to create Being which has meaning and purpose. In its action this freedom does not depend on more or less deliberation. In perfect freedom lies perfect power. A free will is a will which is perfectly one with itself, untouched by any contrary tendencies[2]. God is therefore beyond the region of praise or blame. His only norm of action is the perfection of His own nature, from which He can never deviate[3].

In spite of its perfect union with the *νοῦς*, Cudworth does not hesitate to attribute to the Will of God a sphere of contingent action in a morally indifferent choice. He sees in this nothing contradictory to the perfect dependence on the *νοῦς* since the whole importance of this dependence lies in the moral "determination" of the Will by the *νοῦς*, which is not at stake in contingent choice. A nice example of such a contingent choice is afforded us in the creation of the world, when God had to decide the actual size of the universe, or whether there were to be an even or an odd number of stars[4].

[1] T. I. S. 564. There being no created being essentially good and wise, but all by participation.

Ibid. 737. (The measure of truth and the rule of the will lie in the transcendence.)

[2] Muirh. Manuscr. 69. Howbeit, God may be said in a pure and refined sense to be most perfectly *αὐτεξούσιος*, SUI DOMINUS, Lord and master of himself, and most truly free not in a compound manner as if there were such a duplicity in him as in free-willed beings, or as if he were so much distant from himself and hung so loose together, but as being essentially simple goodness, the root of all wisdom, excellency and perfection.

Manuscript 4980 p. 30. This power of free-willed doing which they have over themselves for good, is but an imperfect power, for perfect power over oneself is no other than essential goodness and wisdom, where the reduplication of free-willed beings ceases and is swallowed up in the simplicity of absolute perfection.

[3] Fr. W. 17. ... For God hath no law but the perfection of his own nature.

[4] Ibid. 16–7.

Ibid. 53. Notwithstanding which, arbitrary and contingent liberty is not quite excluded from the Deity by us, there being many cases in which there is no best, but a great scope and latitude for things to be determined either this way, or that

Calvin erroneously enlarged this sphere of contingency in the Will of God to morally different choice, and was thereby not able to escape the formidable consequence of declaring God as the origin of evil[1].

It is difficult to reconcile these two predications of the divine Will. How can a thoroughly rational will be considered to have, as it were, an a-rational field of action? It could only be on the supposition of an uncritical separating and contrasting Will against Wisdom in God, from which mutual relativation would inevitably follow.

True liberty of will is found only in a will which is perfectly single. A will, torn in different directions at once, is necessarily weakened, since the differing tendencies restrict each other, and prevent the will from acting in the fullness of its power. Such a will is therefore never entirely free[2].

The definition of the perfect Will gives the pattern for the definition of the finite will as the "restless desire for the supreme Good", able to act only towards the Perfect. Cudworth unperturbably clings to the Socratic paradox that no one sins by free will[3]. *(οὐδεὶς ἑκὼν ἁμαρτάνει)*

way, by the arbitrary will and pleasure of God Almighty. As for instance the world being supposed to be finite, ... that it should be just of such a bigness, and not a jot less or bigger, is by the arbitrary appointment of God, since no man can with reason affirm that it was absolutely best that it should have been not so much as an inch or hair's breadth bigger or lesser than it is.

... So likewise the number of created angels and human souls, or that everyone of us had a being and a consciousness of ourselves, must needs be determined by the arbitrary will and pleasure of the Deity, who can obliterate and blot any one of us out again out of being, and yet the world not be a jot the less perfect by it.

[1] Ibid. 78. Moreover it is certain that God cannot determine and decree all human volitions and actions, but that he must be the sole cause of all the sin and moral evil in it, and men be totally free from the guilt of them.

But in truth this will destroy the reality of moral good and evil, virtue and vice, and make them nothing but mere names and mockeries.

[2] Ibid. 48. A perfect being can neither be more nor less in intention, being a pure act it can have no such thing as self recollection, vigilance, circumspection or diligence in execution, but it is immutable or unchangeable goodness, and wisdom undefectible.

[3] Fr. W. 33. Nevertheless they themselves (the herd of modern philosophers and theologers, who zealously maintain the phenomenon of liberum arbitrium) acknowledge that there is so much of necessary nature even in this blind and fortuitous will, that it is notwithstanding always determined to good, or some appearance of it, and can never possibly choose evil when represented to it by the understanding as wholly

Experience of ourselves testifies that we have, and are conscious of having, a certain freedom of choice, or we should not spontaneously make ourselves and others responsible for our actions and theirs. We actually consider ourselves and others, before God and man, worthy and capable of receiving reward and punishment for our deeds.

Freedom of choice presupposes a broken will with differing tendencies. It is therefore the specific condition of imperfect beings[1]. Freedom of choice is neither a distinction, nor a perfection, in the sense that it can be attributed to God; it is a feature of finite beings only. To them it is proper to exist in the realm of "Between" in every respect[2]: between freedom and determination, between perfect knowledge and perfect ignorance, between absolute goodness and absolute evil. Absolute goodness would be a state where every evil is overcome and transformed into good. Absolute evil, however, could not be experienced as evil, because it would be no longer opposed to good. Problems arise only in the sphere of the "Between". The "Between" is the proper wrestling place of philosophy.

Freedom of will as such is not a perfection, it is relative and far more restricted than we are commonly inclined to think. A condition of ab-

such. But within that latitude and compass of apparent good in the understanding, the will to them is free to determine itself to either greater or lesser.

2. Sermon 226. For sin owes its original to nothing else but ignorance and darkness...

For truth has a cognation with the soul...

Plato, Gorgias 499*e*.

[1] Fr. W. 47. This *αὐτεξούσιον* ... which is the foundation of commendation or blame, praise or dispraise, and the object of retributive justice ... rewarding or punishing, is not a pure perfection (as many boast it to be) but hath a mixture of imperfection in it.

Ibid. 63–4. For what is more common than in writings both ancient and modern, to find men creaking and boasting of the *ἐξουσία τῶν ἀντικειμένων* THE LIBERTY OF CONTRARIETY, i. e. to good and evil, as if this was really a liberty of perfection, to be in a indifferent equilibrious state to do good or evil moral, which is too like the language of the first tempter, thou shalt be as God knowing good and evil.

[2] Ibid. 45. Were the soul necessarily and essentially good and impeccable, it would be above this self power, were it nothing but lust and HORME, it would be below it. Now it is in a middle state a perfection betwixt both.

Muirh. Manuscr. 67. ... All freewilled beings which are neither essentially good nor bad...

solute indetermination is inconceivable. Every action is determined at least by its consequences; also its modality changes with the progress of time; that which may seem contingent in the future can already be determined in the present and is certainly necessary as passed.

Cudworth illustrates the broken will by distinguishing higher and lower faculties in the soul[1]. Both of them are, as such, morally indifferent, yet the higher faculties have a natural imperium over the lower. If this imperium is realised, then both, higher and lower faculties, collaborate towards the supreme Good[2]. But when the order is reversed and the lower faculties reign over the higher, the soul loses, not only the higher, but also the lower faculties and perishes. Because the soul has, to a certain extent, the power to realise this natural order, it has within itself the capacity of perfecting or destroying itself[3]. When it is guided by the higher faculties it transcends itself by transcending its lower faculties[4]. By exerting this natural imperium, the soul does not destroy the lower faculties, rather it leads them to health and strength, and turns all the capabilities of doing evil into good. Thus the soul

[1] T. I. S. 220. There is a necessity that there should be higher and lower inclinations in rational beings vitally united to bodies.

[2] Fr. W. 30–1.

2. Sermon 231. Why should it seem strange that the superiour faculties of the soul should become predominant, since they are of a lordly nature, and made to rule, and the inferiour faculties of a servile temper, and made to be subject.

[3] T. I. S. 221. Rational creatures being by means thereof, in a capability of acting contrary to God's will and law, as well as their own true nature and good; and other things kindred of that perfection, which the divine goodness would else have imparted to them.

Muirh. Manuscr. 67. It (the soul) may be more or less watchful and careful and circumspect, excite or quicken itself more or less to the use of endeavours, fortify itself by resolution, wake considerations both rational and fantastical, more or less to confirm its purposes. It hath plainly in it a power or strength to promote itself towards the higher good.

When it (the soul) is recollected into itself and stands upon its watch it will immediately repel those assaults and temptations, which at another time it would be easily vulnerable by.

[4] Fr. W. 19. He that resisting these lower and worser inclinations, firmly adhereth to the better principle or dictate of honesty and virtue, hath in all ages and places in the world been accounted *ἐπαίνετος*, praiseworthy, as being *κρείττων ἑαυτοῦ* superior to himself, or a self-conqueror.

can, by faithful exercise, gradually attain to a fixation in good. But when its lower powers are given dominion, the consequence is that they weaken and finally destroy themselves and extinguish the higher ones. Thus a soul can partly perish and die, even while its body is still alive[1]. Cudworth described the great responsibility which freewill puts upon us in the following way. Whether we fix ourselves in this life towards life or death will have an eternal significance after death, because the health of our soul rests with our deeds. There is no place for contingency whatever in the moral sphere, since, in all our actions, we stand directly before the face of the Absolute, bearing our full share of responsibility[2].

Cudworth believes in a higher life, viz. that of angels who have attained to that final "fixation in good", which to us mortals can never be given. Our human condition remains precarious and uncertain, because the inclination to evil is in our own will. We are therefore always fully exposed to its menace, and have no place of safety within us.

Here we can clearly see, how much better it corresponds with Cudworth's true thought, to see evil in the soul itself, rather than in dead matter[3].

[1] Fr. W. 42–3. But this is not one single battle or combat only, but commonly a long lasting or continued war and colluctation betwixt the higher and the lower principle, in which there are many vicissitudes, reciprocations, and alternations upward and downward, as in the scales of a pair of balances, before there come to be a perfect conquest on either side, or fixation and settling of the soul either in the better or the worse.

[2] 1. Sermon 51–2. All sinne is direct rebellion against God.

... God can never smile upon it...

God and sin can never agree together.

Cf. Plato, Nom. 903.

[3] (Here also we find Cudworth's dogmatic tendency side by side with the critical.

As elsewhere he derives evil from corporeal dispositions alien to the soul, so he again keeps to the doctrine that the soul had originally been created as a harmony, which is now disturbed from outside by sin, but which can be restored by God, the great HARMOSTES.)

2. Sermon 227. For sin is but a stranger and foreiner in the soul, an usurper and intruder into the Lord's inheritance. Sin is no nature..., but an adventitious and extraneous thing; and the true and ancient nature of the soul of man suffers violence under it, and is oppressed by it. It is nothing else but the preternatural state

Freewill cannot be explained by what Cudworth called the point of view of vulgar philosophy, whose characteristic it is to divide reason and will[1]. This assumption inevitably leads into the blind alley of mutual presupposition. There are two alternatives: *a)* from the fact that will can only strive after that which it knows, the conclusion is commonly drawn that the last judgment of reason decides the choice and determines the will; *b)* yet, on the other hand, we must admit that the will has priority over the intellect, at least in so far as the intellect itself is activated by an act of will. Some would make this priority absolute by assuming that, even after the last judgment of reason, the will is still set before a contingent choice entirely uninfluenced by the intellect. Such a will, however, would be a mere impetus, and as such neither true nor false, infallible one could say; yet there would be found in it no continuity in moral habits, nor moral development, nor forming of habits, no reasonable setting and pursuing of purposes. Mere arbitrariness destructive to the very essence of the will would be the result.

Will and intellect presuppose each other. Will is primary to intellect in so far as the act of thinking originates in will; and vice versa, intellect is primary to the will in so far, as will depends on the judgment of reason for its decision.

Cudworth finds the solution to this difficulty again in the unity of the Trinity, where there is no such division between intellect and will. Yet the problem is not solved either, if intellect and will are assumed to be absolutely identical.

In the finite world the unity of the Trinity has its image in the *αὐτεξούσιον* which is the agent of both, will and intellect[2]. The difficulty of mutual presupposition necessarily follows, if intellect and will are con-

of rational beings, and therefore we have no reason to think it must needs be perpetual and unalterable.

Ibid. 227–8. The soul of man was harmonical as God at first made it till sin, disordering the strings and faculties, put it out of tune and marred the musick of it.

(Later it will become still clearer, how sharply such an assumption is opposed to Cudworth's genuine way of thought namely that virtually evil is already inherent in the nature of finite beings.)

[1] Fr. W. 20 f.

[2] Ibid. 25–6. And thus may it well be conceived that one and the same reasonable soul in us may both will understandingly or knowingly of what it wills; and understand or think of this or that object willingly.

sidered as two independent agents, while they are in truth only manifestations of one agent which stands above both and activates both[1]. The assumption of "absolute reason" and "absolute will" contradicts our experience. Reason, if it were an abstract, independent agent would necessarily proceed in an uninterrupted progress of necessary connections of thought. There would be no room for an outside world; reason therefore would be incapable of stimulating the will, because it could not conceive any practical task at all.

Will, on the other hand, if it were an absolute agent, would be blind, a mere impetus which as such is also incapable of action. From the experience that our reason and will are both capable of acting, we are led to assume an agent within us which activates both, reason and will, and which can therefore never be itself direct object of knowledge. The soul, whenever it is recollected upon itself, reflecting on the content of its consciousness, breaks into two parts, a thinking subject and an object of knowledge. In this falling apart it is, as far as object of knowledge, "outside world" and alien to itself.

Already in the examination of knowledge we were led to an indirect conclusion of the *αὐτεξούσιον*, from the nature of sense perception and thought, and also from the fact of self-consciousness. We are again forced to this same inference by the examination of will, to which we will now turn our attention.

A contingent choice could never be decided, if the soul had not a self-determining power [2]. Yet experience shows that we actually are capable of it. The contingency can lie either in the object or in the subject. It is in the object, when the things to be chosen are absolutely alike; in the subject, when this is itself indifferent to any kind of proposed objects. If the determining cause did not lie in the soul itself, the decision of one and the same contingent choice would necessarily always be the same. But experience shows the contrary. This proves that the decision is made by the subject; and the act of decision can be explained only by assuming, that the soul "puts something of its own

[1] Ibid. 24.

[2] Ibid. 15. From hence, alone, it appears that rational beings, or human souls, can extend themselves further than they suffer, that they can actively change themselves and determine themselves contingently and fortuitously, when they are not necessarily determined by causes antecedent.

into the scales[1]", deciding thus by self-determination the indifferent choice presented by reason.

Also animals have this faculty of deciding a contingent choice[2]. Buridan's donkey would not starve to death between two similar heaps of hay through being unable to decide, from which pile to eat. A power of acting upon themselves must therefore be attributed to animals also, but this implies that they have a certain self-consciousness, though in the widest sense. Cudworth finds this hypothesis confirmed by the fact that animals can be domesticated and trained.

The decision of any contingent choice is utterly inexplicable upon merely mechanical principles, because it implies a self-activity of the soul. Thereby Cudworth probably pointed to the most conclusive argument against the Cartesian hypothesis that animals were possibly moved only by mechanical force.

We have already mentioned the assumption of a sphere of contingent choice even in God. Matter alone entirely lacks this faculty and abides always in complete passivity[3].

True liberty of will in finite beings does not lie, however, in the faculty of self-determination in a contingent choice, but rather in the capacity of acting in a morally different choice, between better and worse. We are able to act even while we are still in doubt about the value of the things put before our choice. The will can act, before reason has reached a conclusive judgment[4]. But that means that in will we can transcend reason; and for our practical life this is indeed of vital importance,

[1] Ibid. 13. ... Rational creatures, can add or cast in something of their own to turn the scales when even...

Ibid. 15–6. (In contingent choice.) Here, therefore, is a sufficient cause which is not necessary, here is some thing changing itself, and acting upon itself, a thing which, though indifferent as to reason, yet can determine itself and take away that passive indifference.

[2] Ibid. 56. Now it is not easy to exclude brute animals from such a contingency as this...

[3] Ibid. 55.

[4] Ibid. 39–40. That is when we have no clear and distinct conception of the truth of a proposition (which is the knowledge of it and can never be false) we may notwithstanding, extend our assents further and judge stochastically, that is opine, this way or that way concerning it, and that sometimes with a great deal of confidence and assurance too.

since, as we have seen, we never attain to evidence in the realm of empirical knowledge. If we could act only upon evident knowledge, we could never act at all and should be utterly at a loss in managing the affairs of our daily life[1].

Here we meet once more with the uncritical bent in Cudworth's thought, when he asserts that in our will we can transcend even a clear and distinct judgment[2]. This assertion leads to difficulties indeed. It may be, however, that Cudworth had in mind his practical "solution" of the problem of evidence: that in our inadequate knowledge of facts we are justified in trusting to a certain clearness of apprehension. In this case "clear and distinct judgment" would mean no more than apprehension with a certain conclusive "evidence".

Or does he assume transcendency in the will over a necessary judgment only as an abstract possibility which can never occur in practical life?

Cudworth denies that the will, even after the last judgment of reason, finds itself still before a contingent choice. If this were true, will would be an irrational impetus contrary to its own essence, which is its being directed towards Truth. Yet Cudworth pleads this same hypothesis for the *αὐτεξούσιον*, which is the agent of the will, in connection with the experimental fact that even a necessary judgment of reason differs in its effect upon the thinking subject, according to his actual moral disposition.

But where is now the connecting link between this assertion and Cudworth's theory of logical and moral error? Cudworth firmly clings to the Socratic paradox that nobody sins by freewill. Error and sin

[1] Fr. W. 39. Besides which, it is certain, that in our practical judgments we often extend ourselves or assent further than our understanding as necessary nature goes; that is, further than our clear and distinct perceptions.

(I give the following interpretation not without hesitation. I cannot say with certainty, whether or not, Cudworth really thought it possible that we can transcend in the will even a necessary judgment of reason. The difficulty is that Cudworth, against his own principles, again and again speaks of clear and distinct judgments.

On the other hand, precisely this passage could be taken as a proof of his critical tendency, that knowledge of mere being is not yet knowledge in the true sense.

This again shows how tightly twisted together in Cudworth's thought are the two tendencies.)

[2] Fr. W. 38.

are not directly explicable either from reason[1] nor from will, because this would ultimately mean, that God were their *αἰτία*, since every act of reason and will is a direct participation in divine wisdom and power. Error and sin have their origin in the *αὐτεξούσιον*, which has a certain freedom of choice in deciding to act after more or less deliberation[2]. Solely in this choice lies the *ἐφ' ἡμῖν* which makes us responsible for all our actions. This is another argument against assuming free choice in God, since in the pure act of the divine mind there certainly cannot be more or less of deliberation.

In this theory of knowledge, at least as far as he followed his critical bent, Cudworth shows that we never attain to absolutely conclusive knowledge of facts, and therefore there cannot be any action which is perfectly fixed in the Good[3]. Nowhere are we free from risk and peril. And, even if we could attain to clear and distinct knowledge, practical life often does not leave us enough time, but urges us to action before any definite result of deliberation is reached.

If Cudworth sees the liberty of will and the source of error solely in the choice of more or less deliberation, he indirectly affirms, in agree-

[1] Ibid. 35. Because reason, as such, can never act unreasonably, understanding, as such, and clear perceptions, can never err. There is no such thing as false knowledge nor erroneous understanding, nor can sin ever be the result of reason ... any more than error. Nor is error any more from God and the necessary nature of understanding, than sin is.

[2] Ibid. 37. A man's soul as hegemonical over itself, having a power of intending and exerting itself more or less in consideration and deliberation, when different objects, or ends, or mediums, are propounded to his choice, that are themselves really better and worse, may, upon slight considerations and immature deliberations (he attending to some appearance of good in one of them without taking notice of the evils attending it), choose and prefer that which is really worse before the better, so as to deserve blame thereby.

But this not because it had by nature an equal indifferency and freedom to a greater or lesser good, which is absurd, or because it had a natural liberty of will either to follow or not follow its own last practical judgment, which is all one as to say a liberty to follow or not its own volition.

But because he might have made a better judgment than now he did, had he more intensely considered, and more maturely deliberated, which, that he did not, was his own fault.

[3] Muirh. Manuscr. 66. While it is true that the idea of "good" or what is "in congruity with the soul", is always present, it is only so as a "general and confused notion".

ment with the Socratic paradox, that the *αὐτεξούσιον* actually would act in accordance with the necessary judgments of reason. With this, a possible transcendence of a necessary judgment of reason seems inconsistent.

Not only the actual degree of knowledge and the goodness of action depends on the *αὐτεξούσιον*, but, what is far more, even the very faculty of knowledge and will; knowledge and will are either strengthened or weakened by the acquiring of their respective habits[1]. It is this which makes us the object of praise or blame before God and men and which determines the condition of the soul, not only in this life, but in the future life also. Considering, how much depends on the power of, and will to, knowledge, in our souls, we understand, why Cudworth so emphatically calls us to unrelenting watchfulness of mind, reminding us of the urgent warning of Plato, that, despite all our uncertainty in every respect, the one thing we must at all costs avoid is to become *μισόλογοι*, haters of the LOGOS)[2]. For, if we were to yield to this temptation of despondency, we finally and for ever should cut ourselves off from all Truth, and, what is worse, deprive ourselves of all hope of ever finding Truth[3].

Yet even with the hypothesis of the *αὐτεξούσιον* we do not escape at least a certain mutual presupposition of reason and will. For, through the training of our cognoscitive powers, the *αὐτεξούσιον* is strengthened in the will, which, in its turn, stimulates to new efforts of thinking. In other words, only as far as the soul succeeds in transcending its own finiteness, and, by an effort of will, rises up to the universality of the ideas, can it attain to knowledge. And, on the other hand, the purer the knowledge is, the purer also will the action be. This inextricable relationship of logic and ethics can probably never be overcome, since its ultimate reason is the unity of the Trinity.

Beyond the problem of freewill, which Cudworth so keenly felt that he had not solved[4], still wider horizons open up before his mind, in

[1] Plato, Nom. 904*a*.

[2] Plato, Phaedo 89*c*.

[3] Plato, Phaedo, 90*d*.

[4] Muirh. Manuscr. 63. Hence, if what I shall say concerning free will seem unsatisfactory to any, I shall think it no marvel at all, for I never was myself satisfied in any discourse which I read of it.

which, once more, there is seen something of the natural imperium of ethics over logic. Cudworth believes, that, despite many distressing limitations, even finite Beings can still go some of the way towards true freedom, which lies beyond the freedom of choice. He conceives, as we have seen, the dependence of the Will of God on His supreme Goodness, both as direct and as mediated by the *νοῦς*. In analogy to this, he believes in a faculty in man of a higher participation in God, which is not, in the strict sense, mediated by the *νοῦς*[1]. Within the compass of our finite life, we are able to widen our own will to the measure of the divine will, so much so that it can happen that it is God Himself Who acts in us[2]. This "entering into the divine life", where God "is

[1] 1. Sermon 16–7. Whereas every true christian finds the least dram of hearty affection towards God, to be more cordiall and sovereign to his soul than all the speculative notions, and opinions in the world ... He feeleth himself safely anchored in God, ... neither is he scared with those childish affrightments, with which some would force their private conceits upon him; he is above the superstitious dreading of mere speculative opinions; as well as the superstitious reverence of outward ceremonies: he cares not so much for subtlety, as for soundnesse and health of mind.

Ibid. 19. But it is a piece of that corruption that runneth through humane nature, that we naturally prize truth, more than goodnesse; knowledge, more than holinesse. We think it a gallant thing to be fluttering up to heaven with our wings of knowledge and speculation, whereas the highest mystery of a divine life here, and of perfect happinesse hereafter, consisteth in nothing but mere obedience to the divine will.

Happinesse is nothing but that inward sweet delight, that will arise from the harmonious agreement between our wills and God's will.

[2] ... Ibid. Pref. The scope of this sermon ... was not to contend for this or that opinion, but onely to perswade men to the life of Christ, as the pith and kernel of all religion. Without which, I may boldly say, all the severall forms of religion in the world, though we please our selves never so much in them, are but so many severall dreams. And those many opinions about religion, that are every where so eagerly contended for on all sides, where this doth not lie at the bottome, are but so many shadows fighting with one another.

Ibid. 33–4. Nay, this divine life begun and kindled in any heart, wheresoever it be, is something of God in flesh; and, in a sober and qualified sense, Divinity incarnate; and all particular christians, that are really possessed of it, so many mysticall Christs.

And, God forbid, that God's own life and nature here in the world, should be forlorn, forsaken and abandoned, of God himself...

Never was any tender infant, so dear to those bowels that begat it, as an infant new-born Christ, formed in the heart of any true believer, to God the father of it ...

born in our hearts", is Love, and Love directly includes Truth, freedom[1], and beatitude[2].

This widening of the finite will to the Will of God is the health of the soul, because thus it overcomes, to a certain extent, the rupture in the will, which is its deficiency[3]. It thereby also gains a fuller power of knowledge[4]. This overcoming of the broken will is not given to us as a

the life of God in us; which is nothing else but God's own self communicated to us, his own Sonne born in our hearts.

Ibid. 78–9. Whosoever is once acquainted with this disposition of spirit, he never desires any thing else: and he loves the life of God in himself, dearer than his own life.

Plato, Res Publ. 613*a*.

[1] 2. Sermon 229–30. For the spirit is not always to be taken for a breath or impulse from without but, also, for an inward propension of the soul; awakened and revived in it, to return to its proper state, as it is intellectual, and then to act freely in it, according to its ancient nature.

For if the spirit were a mere external force acting upon the soul without the concurrence of an innate principle, then to be acted by the spirit would be a state of violence to the soul, which it could not delight always to continue under; whereas the state of spirit is a state of freedome... It is the soul's acting from an inward spring and principle of its own intellectual nature...

[2] 1. Sermon 49. Happinesse is nothing, but the releasing and unfettering of our souls, from all these narrow, scant, and particular good things; and the espousing of them to the highest and most universall good ... which is ... goodnesse it self and this is the same thing that we call holinesse; in it self the most noble, heroicall and generous thing in the world ... God stamped, and printed upon the soul.

Ibid. 60. O divine love ... the joy of God's own heart.

Ibid. 72–3. And as for heaven, we onely gaze abroad, expecting that it should come into us from without, but never look for the beginnings of it to arise within, in our own hearts.

[3] Fr. W. 64. Whereas the true liberty of a man, as it speaks pure perfection, is when by the right use of the faculty of freewill, together with the assistences of divine grace, he is habitually fixed in moral good, or such a state of mind as that he doth freely, readily and easily comply with the law of the divine life, taking a pleasure in complacence thereunto, and having an aversation to the contrary.

Ibid. 65. Whereas true liberty, which is a state of virtue, holiness and righteousness (a communicated divine perfection or participation of the divine nature) can never be abused.

[4] 1. Sermon 81. That is the earthinesse of mens affections, that darkens the eye of their understandings in spirituall things. Truth is alwayes ready, and near at hand, if our eyes were not closed up with mud, that we could but open them, to look

"spark", neither is it a static "fixation in good", to which we can attain once for all; but in an unrelenting effort of self-transcendency it has to be won over and over again. Yet faithful exercise strengthens the soul, and its way becomes less burdensome, the further it advances in perfection. Yet, at the same time, the way becomes harder, because the more the soul lives by the rule of Love, the stronger grows its power of knowledge, and the clearer the comprehension of its own deficiency.

When a soul loves God, it has a direct certainty of His existence, and there opens up a higher possibility of knowledge. The decisive deficiency of finite knowledge in discursive reasoning is the falling apart of subject and object; but even this, in a certain measure, is overcome by a soul which lives in the vision of divine Love. Thus an entirely new possibility of evidence and certainty of Truth comes into sight. Here once more it is seen that Love and Truth are inseparably one. They are the divine Life itself within us. So far as we live in it, our own finiteness and corruptibility are overcome. Love and Truth are therefore the highest reality and power which no enemy can defeat[1]. We have, in fact, no enemy but our broken will; and in so far as this is overcome, no danger is left, and no reason for any more fear, whatever may happen outside[2]. For "to him who is good, nothing is evil"[3].

Cudworth conceives the origin of evil in the necessity *(ἀνάγκη)* of finite existence of all created beings[4], which means that it is essential

upon it. Truth, alwayes waits upon our souls, and offers it self freely to us, as the sun offers its beams, to every eye, that will but open, and let them shine in upon it.

Ibid. Pref. Nay, all true knowledge, doth of it self tend to God, who is the fountain of it; and would ever be raising of our souls up, upon its wings thither, did not we *κατέχειν ἐν ἀδικίᾳ*, detain it and hold it down, in unrighteousnesse.

[1] Ibid. 62. Let us ... follow truth in love ... Truth and love, are the two most powerfull things in the world, and when they both go together they cannot easily be withstood.

[2] Manuscr. 4980, p. 30, see p. 137, note 2.

[3] 1. Sermon 51. Every true saint, carrieth his heaven about with him, in his own heart, ... he might safely wade through hell it self.

T. I. S. 658–9. It being certain, that none are less solicitous concerning such events, than they who are most truly religious. The reason thereof is, because these place their chief good, in nothing that is ... exposed to the strokes of fortune; but in that which is most truly their own, namely the right use of their own will.

[4] Fr. W. 63. Wherefore this *αὐτεξούσιον*, SUI POTESTAS, self power, commonly called liberty of will, is no arbitrary contrivance, or appointment of Deity,

to every finite being to exist in the μεταξύ, in the Between of Being and Not-being, of knowledge and ignorance, of good and evil, of past and future. As Plotinus put it, πέφυκε γὰρ ἐπ' ἄμφω. (Enn. I, 2, 4).

This existence in the Between (μεταξύ) is a constant transition. It is the necessary mode of life for all finite beings. No perfect power could have endowed finite beings with the mode of existence of the Perfect. It is conceivable, that, by violence, and at the cost of freewill, actual sin could have been prevented[1], but such an accidental act of divine power would not have healed the breach which goes through our whole being. Evil dwells not in God, but in ourselves. Therefore it can only be overcome "from within". The overcoming of the broken will can only be achieved by perfect goodness, which takes up evil and within itself transforms it into good[2]. Law, which opposes evil to good as its enemy, and calls us to exclude evil and flee from it, has no healing power; on the contrary, it makes the breach deeper. A forensic justification by God, or any outward ascribing of justice by God, without sanctification, are also incapable of overcoming evil[3]. At best, they

merely by will annexed to rational creatures, but a thing which of necessity belongs to the idea or nature of an imperfect rational being. Whereas perfect being, essentially good and wise, is above this freewill or self-power, it being impossible that it should ever improve itself, much less impair itself.

Manuscr. 4980, p. 147. This is the true accompt of the originall of sin, that it is neither caused by God, nor by any positive substantiall principle, but the possibility thereof proceeds only from the imperfection and deformibility of creatures; but the actual cause of it is never any other than the rationall creature itself, not putting forth that exertive power, which it hath towards the higher principle in its nature, but by sluggish remission and relaxation, sinking down into the lower. It is a great mistake of those that think there is as much positive activity of freewill to evil, as there is to good, which proceeds from a wrong notion of the liberty of will.

Plato, Lysis 218 *a–b*.

[1] T. I. S. 889. For peccability arises from the necessity of imperfect freewilled beings, left to themselves, and therefore could not by omnipotence it self have been excluded; and though sin actual might perhaps have been kept out by force and violence; yet all things computed, it was doubtless most for the good of the whole, that it should not be thus forcibly hindered.

[2] Ibid. 221. (The highest is to turn evil into good.)

[3] 2. Sermon 208. (To believe that there is no evil in sin except the punishment which might follow...) which is to destroy the nature and reality of sin, and to make it nothing but a mere name or phancy.

Ibid. 209. But if sin be not a mere name or phancy, but that which hath a real and intrinsecal evil in it, greater than that of outward punishment; then certainly,

could deliver from exterior punishment. But the real punishment is evil itself, which by deepening the breach in the will, makes us increasingly impotent, miserable and poor [1]. Outward release from punishment can bring no peace as long as the will is divided, for "the corruption of the heart is more sad than the guilt of sin [2]". Again we find two different voices in Cudworth. As a dogmatic theologian, he deems outward, eternal punishment of the wicked to be a central part of the Christian Faith, because therewith "the justice of God is being fulfilled [3]". Yet in his critical and truest thought, he does not fail to see that any absolute confronting of evil with the justice of God would make divine Love itself relative.

The finite stands not only in the formidable Between of good and evil, but also in the glorious Between of the Absolute as "terminus a quo" and the Absolute as "terminus ad quem"; and in this Between lies all our hope, for it means that no creature can ever fall out of the Absolute, which is Perfect Love [4].

Not once for all, yet step by step, can we already, in our life on earth, overcome our broken will by Love.

it cannot be so transcendent a happiness as some men carnally conceit, to have an impunity in sinning to all eternity.

[1] 1. Sermon 36. No surely, there is a weaknesse and impotency in all evil, and masculine strength and vigour in all goodnesse.

2. Sermon 228. Sin is but a disease and dyscrasy in the soul, righteousness is its health and natural complexion of it; and there is a propension in the nature of every thing to return to its proper state, and to cast off whatever is heterogeneous to it.

[2] Fr. W. 30.

[3] Ibid. 4.

T. I. S. 879–880. ... And will at last make it appear, that a thread of exact justice did run through all, and that rewards and punishments are measured out in geometrical proportion.

Cf. ibid. 880. Lastly, it is in it self fit, that there should be some where, a doubtful and cloudy state of things, for the better exercise of vertue and faith. ... Were there no such difficulties to encounter with, no puzles and entanglements of things, no temptations and tryals to assault us; vertue would grow languid.

[4] Ibid. 464.

CONCLUSION

We have come now to the end of our journey through the universe of Cudworth's philosophy. The impression of an overwhelming complexity has resolved itself in the discovery of two disparate currents of thought. This does not mean that Cudworth had adopted the thoughts of others alien to his own and, without properly thinking them through, had given them a place in his philosophy. The reason lies deeper. The critical and dogmatical tendencies are rooted in Cudworth's own personality. He bore within himself in an exceptional way the tremendous tensions of his century. The astonishing and most fascinating thing is that the modern critical philosophy passes into his dogmatical line, which as a Platonist he overcomes and refutes again and again by fresh impetus of thought.

The question now remains whether, and how far, his uncritical tendency was influenced by Cartesius, and in what way his critical thought was inspired by Plato.

Cartesian influence. Cudworth did not admit that Descartes had found a new start in philosophy. As far as he acknowledged his system, he considered it to be a revival of the ancient Pythagorean atomism. But the three metaphysical certainties which Descartes considered as the sure foundation of knowledge and the turning point in philosophy, Cudworth did not consider a possible starting point.

Cudworth agrees with Descartes on two points: in the theory of substances, and in the inference by reason of the outside world. We must however ask how his own theory of substances, disconnected as it is from the Cartesian foundation of knowledge, can still accord with that of Descartes. A second question is how far the inference of reason of the outside world is overcome by his Platonism.

We will consider the first question: the theory of substances. The significance of Cogitation and Extension within the Cartesian system is so entirely different from Cudworth's scheme, that it seems doubtful whether the two theories can be compared at all on one and the same plane. Descartes, as a Geometrician, conceived the two substances as comparable to parallel lines, both indefinite in beginning and end, yet finite since relative to one another. The two parallels meet in the infinite substance, in God. The vital question how any connection between Cogitation and Extension can exist, how Cogitation can think Extension, remains ultimately unanswered, unless by reference to the infinite substance in which both meet. The decisive difference of Cudworth's theory is found in the relation between the substances. Cudworth claims the absolute imperium of Cogitation over Extension. The consequence is that the Final Cause is given a central importance and that Space is finally subordinated to Time. Time is given direction and import as the form of actualisation of finite souls. Even God is conceived of as acting in time when he directs his creative power "ad extra". With this the progress of knowledge after the method of geometrical static proportion is inconsistent. Proportion implies simultaneousness and is the only possible method which corresponds to the co-ordination of Cogitation and Extension. In Cudworth's philosophy there is, strictly speaking, no place at all for a duality of substances.

Secondly, the inference by reason of the outside world. Platonism and Cartesianism meet in Cudworth's philosophy in the problem of whether we are justified dogmatically to assert any reality outside of Cogitation; though not in the sense of a naive realism, yet as mediated by an inference according to reason. The inference is drawn from the clear and distinct intelligibility of Extension and its modifications. Cudworth assumes the Cartesian view against Plato in order to escape the consequence of being forced to deliver up the sensible world to final unintelligibility. He was, however, anxious to say that he regarded it as a mere explanatory hypothesis.

Yet even as a hypothesis the inference according to reason cannot be retained when considered in the light of Cudworth's own theory of knowledge, according to which we can attain to knowledge of Being only in so far as Being can be comprehended by thought. Sensible things are intelligible only in so far as they represent the ideas and are themselves "ideal". Cudworth goes further and, with Plato, identifies Being and Idea. This however does not imply that the sensible world does not exist, but it means that all sensible things are a mixture of Being and Not-being, as they are also a mixture of rest and motion. In so far as the sensible world IS, it is intelligible; and as far as it IS-NOT, it cannot be known. We can therefore never tell what we are really looking at in sense perception, because one part of that which we perceive is withdrawn from our perceptive faculty. Cudworth hoped, by the hypothesis of atomistic physiology to escape this consequence from which Plato did not shrink.

Only as far as the sensible world is intelligible does it exist. This excludes precisely Being outside of Idea; nothing clearly and distinctly intelligible can be thought outside of Idea.

The inference according to reason of reality outside of Idea becomes questionable also from another consideration. Cudworth admits that the function of the lowest energy of soul, "ratio immersa", of Plastic Nature and imagination, are never perfectly comprehensible because causality cannot be thought. We cannot get, as it were, behind the act of thinking. We shall therefore never be able to define the relationship of Cogitation to an outside world essentially alien to it. Only within the ideas are we able to discover and comprehend relations. The inference of the reality of the outside world is there necessarily to be discovered as a *νόθος συλλογισμός*.

Cudworth as Platonist. How far is Cudworth guided by Plato in his refutation of Descartes? In the following, I shall not give a summary of Cudworth's critique of Cartesianism, I shall only show a few of the most important points which he has in common with Plato.

1. Cudworth, like Plato, conceived a latitude in the divine. In Platonism we find this latitude in the conception of the *αἰτία*, which on the one hand implies only the One and Perfect *τὸ ἓν ἀγαθόν*, but on the other hand includes also *νοῦς* and *ψυχή*. Cudworth finds this latitude in the divine through the Trinity. The immediate consequence is, for Cudworth as for Plato, that Ethics, Aesthetics, Logic, and Religion are inseparable from each other and are ultimately one.

2. Above and beyond knowledge and Being is the One and Perfect. Knowledge of mere Being cannot therefore be the ultimate purpose of philosophy. Every true knowledge always includes the knowledge of the "how it is good for the things to be".

3. There is a transcendent uncreated Truth as the divine order and purpose of the Whole. In absolute Truth alone is the foundation of all finite knowledge.

4. The relationship of the Absolute to the finite is transcendental. The transcendent idea is the law of Becoming of all finite Being and knowledge. An absolute transcendence would presuppose an independent, sensible world outside of the Absolute, to which the Absolute would necessarily be relative, and this is a contradiction in terms. The transcendental relationship is symbolised by Cudworth as a participation of finite life and thought in the Absolute. God is not a mere spectator who abides in unapproachable transcendency unmindful of his creatures.

5. A direct consequence of this transcendental relationship of the Perfect to the finite is the fundamental significance given to the final cause of Being and knowledge, and also the impossibility of any coordination of Space and Time; not to mention an imperium of Space over Time as Descartes assumed.

The ideas, moreover, cannot be considered as a storehouse of innate conceptions at our command. We are not on the same level with them, but must look up to them.

6. In knowledge we are dependent on the stimulation of experience. The sensible world presents the problems, but reason when it finds the solutions does not return to it. Only a soul recollected within itself and occupied in pure self-activity can attain to any knowledge, even the partial knowledge of sensible things.

7. Since in thinking we depend upon experience, it is impossible for our knowledge to proceed in a linear progression; no system of sciences can therefore ever be built up by the method of deduction. The Whole is never given to us, it cannot become the object, either of knowledge or of doubt. In knowledge we proceed from one problem to another, but we have the possibility, from whatever particular problem we may start, to push through to the centre of all knowledge, the knowledge of the Perfect.

8. Experience progressing with methodical certainty is impossible also because we never attain to any assurance of the integrity of our cognoscitive powers. We cannot precede the act of thinking by a "cri-

tique of reason", nor can we take our starting point in a demonstration of guaranteed cognoscitive powers; but we are forced to proceed by a defective method. The defectiveness of the method consists in that we cannot think, or even doubt, without presupposing Absolute Truth which is yet the ultimate goal of every act of thought. Our cognoscitive powers can prove their integrity only in practice. In every particular problem they are put to the test anew. Yet their capacity is decisively influenced by our moral habitus. We can thus do much indirectly to develop and strengthen them, while at the same time it is in our power utterly to destroy them.

In two important points Cudworth appears to have been held back from following up his own critical premises by his great concern for Cartesian metaphysics.

When in the pure knowledge of the logical and mathematical axioms Cudworth thinks to have found the Archimedial point "Where all knowledge comes to rest", he does not sufficiently take into account that we principally depend on experience and therefore do not know whether new axioms will not be remembered in us in the future, by which all previous knowledge will perforce be modified.

In his great concern to prove the existence of Absolute Truth, Cudworth has not given sufficient attention to the problem of how the ideas are remembered in us; namely, to the fact that when we apply the immutable ideas to the explanation of the sensible world they enter in us into specific relations, by which their logical content is necessarily modified. Moreover, the thinking subject itself is in process of Becoming in every acquisition of knowledge. This process again modifies the content of the ideas. We must therefore acknowledge that absolute immutability of the ideas cannot be postulated, not even for the logical and mathematical axioms as far as these are thought by us.

When Cudworth speaks of a ladder of perfection within the finite, he follows Neo-Platonic ideas rather than Plato. Such a ladder cannot be asserted when ideas are postulated for all Being. For this means that all finite Being is conceived as existing in a direct relationship with the Perfect, in the sight of which a more or less of imperfection has no meaning. It is quite certain that Cudworth, with Plato, did postulate ideas for all Being, even for negative conceptions, though without working out fully the consequences thereof.

Cudworth goes further than Plato when, upon Platonic principles, but perhaps without the Platonic critical reserve, he explicitly interprets the *νοῦς* as act and content, and when he accordingly modifies the conception of the *ἀνάμνησις*; and secondly, when he conceives of holiness as a direct participation in the will of God, not necessarily mediated by the *νοῦς*.

Cudworth's personality. The cleavage which we have found in his thought must have been a feature of his whole personality. It also appears in the crude outspokenness with which he dealt with his opponents as with "men sunk into the bottomless stupidity of atheism and corrupted in the wickedness of their own hearts". Yet at the same time he would with uncommon candour examine the arguments of these "stupid atheists", and with that fine awareness of goodness in any disguise which was his, he would spare no pains to do them justice, so much so that he himself was by some suspected of a hidden atheism.

An even greater discrepancy is found in his doctrine of eternal punishment which, he says, is "claimed by reason and justice", and should not be rendered harmless by being considered only as a passage on the way to salvation. Such assertions are found close beside the most sublime descriptions of divine love, from which Cudworth sees rising a hope so great that no finite spirit can ever conceive it, nor bear the full weight of its expectancy.

A similar divergency underlies the assertion that men are actually given the possibility of corrupting their own souls to such a degree of disintegration that they fall out of the condition of the *μεταξύ*, the Between of good and evil, having reached a state of "fixation in evil", in spite of the fact that the condition of this Between is essential to the nature of all finite existence, and is itself the very source of evil.

Cudworth conceives an overcoming of the condition of the Between in the opposite direction also. A soul on the way of perfection can gradually overcome the natural brokenness of will; by widening and adapting itself more and more to the divine will it can attain already here on earth a certain degree of holiness and fixation in good.

Holiness, as Cudworth sees it, is the overcoming of evil by transforming it into good; and this once for all excludes the existence of any evil which could stand as an absolute entity against good. No evil can be conceived which has fallen out of the possibility of being transformed by Perfect Goodness.

The dogmatic and critical tendencies do not only run like two streams through the whole of Cudworth's thought. They even appear in the outward presentation of his work. There is the broad slow-moving progression of traditional scholastic argument with endless quotations and manifold digressions side by side with the quick, tense and powerful energy of thought expressed in a language full of image and life.

The disproportion of the outward presentation of the T. I. S. stands indeed in startling contrast to the sublime *symmetries* of thought contained within it, wherein we find, perhaps, the truest image of Cudworth's personality. Cudworth's philosophy can be contemplated from the point of view of *τάξις* (order); most conspicuously revealed in his doctrine of the Trinity. On symmetry ad intra and ad extra is founded Being and Truth of every particular thing. Symmetry is the beauty, goodness, and relative perfection of all finite beings. All things have an inner urge towards *τάξις*. Evil, according to Cudworth, is not a nature, but the inversion of *τάξις*. Evil IS not as a fact.

This inversion can occur in every sphere: in knowledge, when thought is derived from sensation; in the theory of Being when matter is put forth as the first principle; in aesthetics when beauty is sought, not as a manifestation of the supreme Good and a stimulation on the way to perfection, but is selfishly enjoyed for its own sake. And in religion, when in the doctrine of forensic justification outward punishment, and not the corruption of the soul, is considered as the evil, from which we desire deliverance. Evil opposes and corrupts Nature in the pre-eminent sense; it is therefore always direct rebellion against God.

Symmetry is the inner law for all finite Being in the process of Becoming. Symmetry actually IS its Being. When Cudworth uses the term "from inside" he is far from meaning an inside world standing against an outside world, but the outside world must be taken back into the centre, for there, in its being founded upon and in the Absolute, it will find its true meaning and purpose, its Order. In no way can this imply depravation and neglect of the sensible world. Cudworth, as a genuine Platonist, naturally thinks from the centre and towards the centre.

Yet it is true that we meet in Cudworth a secret urge for getting beyond the middle sphere of vital union. He sees the soul as created for a purer participation in the divine life than we can attain to here on earth. The soul has a natural urge towards its original purpose and ultimate end.

The term "from inside" is seen in a yet different light in the theory of knowledge, where Cudworth says that "the spirit knows itself by its own"; and in the same way are the values activated in our conscience. In aesthetic contemplation the spirit can perceive beauty only by recreating it in itself. This it can do solely by virtue of its own congruence with beauty.

Such an inner congruity is also the "condicio sine qua non" of faith. Far from being merely belief in a fact beyond understanding, faith is „a divine power in the soul". A soul that lives in faith possesses a certainty of the Absolute by virtue of its own affinity with it. This certainty is experienced in a higher order of evidence. Faith leads therefore straight on to holiness, in which we gain an immediate certainty of God, because we bear his image in our own heart in so far as we act according to his will. "In the meditation of the outward cross we beg for the inward cross."

The pre-eminent significance of the centre in Cudworth's philosophy is also reflected in the conception of the Whole which runs throughout his thought. The universe is one Whole, and as a whole it is good. But also every order of relations, as far as we conceive it, is a unity in multiplicity, through its being directed towards the Final Cause. Reason has the power of gathering up the Many into One. It thus creates one Whole, one *κοινωνία τῶν γενῶν* in every act of thinking. The thinking subject also is one Whole. As a unity in multiplicity, symbolised by the *αὐτεξούσιον*, man activates all his faculties of soul. He thereby becomes an image of the Perfect Mind, who in one pure act knows himself eternally.

We see yet another glimpse of Cudworth's personality and the wonderful candour which must have been his, when we hear him speak of the "immense capacity of pleasure" with which God "who delights in the happiness of his creatures" has endowed us[1].

[1] Manuscr. 4983, p. 94. Lord, how long have we lived, how little have we remembered the immortality of our nature. Custom, education, some transient fears,

All that God does is for joy. He created the world for joy, the joy of angels, men, and animals. Sense perception, the passions of our own soul, far from being given us for deception, are a source of joy for us. A certain freedom in the search for Truth is granted us, that the joy of discovery might be ours. The spirit exults when in beauty, truth, and goodness it beholds its own. In faith the mind rejoices in its immediate certainty of God; and faith is perfected in holiness, the joy of God's own heart in us.

So we see Cudworth's philosophy comparable to a sublimely differentiated geometrical figure whose symmetries are full of beauty and grace. It appears the more beautiful as it is seen against the background of a tremendous vigour of thought and much toilsome work in the going up and going down the first four steps of knowledge, as Plato puts it[1]. This work Cudworth did in the serene atmosphere of the true Platonic *συμφιλοσοφεῖν* with philosophers of all ages and from many lands *ἄνευ φθόνου*, without envy. Whoever does not shrink from the effort to recreate it within himself can hope that the fire which enkindled the Master will pass on to him to light up in his mind that which words cannot express.

ῥητὸν γὰρ οὐδαμῶς.

sudden thought, lead us to thy altar; but how little do we mind the interests of the mind, and we do the interests of this life. How little do we think of Thee, from whom we have received such an immense capacity of pleasure, and how foolishly do we forget the everlasting pleasures of eternity ... If Thou dost us so much good, why are we not filled with the expectation of ineffable satisfaction, that Thou hast so kindly laid before us. Why don't we depose all the foolish imaginations of happinesse and pleasure, which we are every day building in our childish fancies, and we have such excellent things to hope for.

[1] For this and the following see Plato, Epistle vii 342*a*, 344*b*, 341*c*.

CONTENTS